Human Robotics

Human Robotics

Neuromechanics and Motor Control

Etienne Burdet, David W. Franklin, and Theodore E. Milner

The MIT Press
Cambridge, Massachusetts
London, England

MIT Press books may be purchased at special quantity discounts for business or sales promotional use. For information, please email special_sales@mitpress.mit.edu or write to Special Sales Department, The MIT Press, 55 Hayward Street, Cambridge, MA 02142.

This book was set in Syntax and Times by Toppan Best-set Premedia Limited. Printed and bound in the United States of America.

Library of Congress Cataloging-in-Publication Data

Burdet, Etienne, 1965–
Human robotics : neuromechanics and motor control / Etienne Burdet, David W. Franklin, and Theodore E. Milner.
    pages cm
Includes bibliographical references and index.
ISBN 978-0-262-01953-8 (hardcover : alk. paper) 1. Computational neuroscience. 2. Biomechanics. 3. Human mechanics. 4. Robotics. I. Franklin, David W., 1971– II. Milner, Theodore E., 1953– III. Title. IV. Title: Neuromechanics and motor control.
QP357.5.B85 2013
612.8—dc23
2013001550

10   9   8   7   6   5   4   3   2   1

We dedicate this book
    to Mei-Wen, Jacques, and Nicolas,
    to all of my family especially to my parents and my wife, Sae,
    to Silia and Victoria, Ted's rainbow and sunshine.

Contents

Preface

This book provides a comprehensive and rigorous treatment of the control of human movement from the perspectives of both the adaptation of the neural control system and the adaptation of the properties of the mechanical plant, incorporating approaches from physiology, engineering, and computational neuroscience. Failure to consider either neural or mechanical contributions to observed behavior can lead to erroneous conclusions. For instance, there are examples in the literature in which the effects of dynamics and muscle mechanics are wrongly attributed to neural control. Therefore, we and others have developed a synthesis of musculoskeletal biomechanics and neural control over the past thirty years that we call *human robotics*. Why this name? We use the framework of robotics to understand the control problems that are being solved by the human motor system, and the insights gained by this approach lead to more versatile robots with humanlike capabilities.

When we taught this material, we could not find any suitable book treating all of the different aspects in a coordinated manner. Consequently, we created our own lecture notes, which evolved into this book. Ted Milner started a course on this topic at Simon Fraser University twenty years ago. Using this course as a basis, Etienne Burdet and Ted developed tutorials for the 2004 IEEE/RSJ International Conference on Intelligent Robots and Systems. This set of tutorials evolved into lecture notes through the interaction with David Franklin, integrating our collaborative research over the past fifteen years. The notes have now been used for courses at Simon Fraser University and McGill University in Canada, Imperial College London, National University Hospital in Singapore, and Université Pierre et Marie Curie Paris VI, geared to motor control researchers, bioengineers, physiotherapists, and roboticists. Participants in these courses have encouraged us to present the material in the form of a book. The result is this book: *Human Robotics: Neuromechanics and Motor Control*. The authors are listed in alphabetical order, reflecting that the three of us contributed equally to its preparation. Its content has been greatly enriched by discussions with many colleagues and their feedback on our text and suggestions, as well as contributions to figures and text. We would like to thank in particular (in alphabetical order) Alaa Ahmed, Carlo Bagnato, Sivakumar Balasubramanian, Mei-Wen Burdet-Liu,

Domenico Campolo, Themistoklis Charalambous, Sae Franklin, Roger Gassert, Ganesh Gowrishankar, Emmanuel Guigon, Ian Howard, Alejandro Melendez, Eric Perreault, Luc Selen, Atsushi Takagi, and Daniel Wolpert.

A summary of the variables introduced in the book is provided in the appendix. The following conventions are used throughout the book:

- Terms in *italic* generally identify new concepts as they are introduced.

- SI units are used throughout the book.

- In mathematical formulae, the logical $\forall$ means "for all" and $s \in S$ "s is an element of the set S." Furthermore, scalar variables are represented in italic (e.g., x), vectors are represented in bold lowercase (e.g., $\mathbf{v}$), and matrices are represented in bold uppercase (e.g., $\mathbf{M}$). Sets—that is, a collection of abstract objects a_1, a_2, and so on—are represented by $\{a_1, a_2, \ldots\}$. By default, vectors are column vectors and otherwise indicated by the transpose.

This manuscript can be used as a textbook for final-year undergraduate and graduate students in psychology, kinesiology, rehabilitation therapy, neurology, physics, computer science, or engineering and as a reference for professionals in these fields who are interested in understanding the physiology and algorithms of human motor control. It is written in a self-contained manner, such that the material can be understood by these distinct communities with their different perspectives and knowledge. The mathematical background required for an adequate understanding includes introductory undergraduate-level calculus and linear algebra as presented in standard textbooks such as Mathematical Methods for Physics and Engineering (KF Riley, MP Hobson, and SJ Bence, Cambridge University Press, 2006).

Etienne Burdet, David W. Franklin, and Theodore E. Milner
London, Cambridge, Montreal

1 Introduction and Main Concepts

1.1 "Human Robotics" Approach to Model Human Motor Behavior

How do humans perform skillful actions and continually improve performance with such ease? How do children develop complex motor behaviors? How could one best assist persons whose ability to control movement has been impaired by neurological disease or injury? These intriguing questions have driven research on human motor control for more than a hundred years, but much is still unknown about the neural mechanisms involved in the learning and control of motor tasks. Although it may not be possible to develop equations that completely describe the neural control, this book proposes *a constructive approach to motor control* in which models are developed based on evidence from physiological experiments using an approach reminiscent of physics. These computational models yield *algorithms for motor control*, which may then be used as tools to investigate or treat diseases affecting the sensorimotor system as well as to guide the development of efficient computer algorithms and hardware that can be incorporated into new products designed to assist in the tasks of daily living.

Currently, physiologists and rehabilitation scientists often turn to robotics and control theory for quantitative and objective analysis methods, which has led to significant advances in understanding human motor control over the past thirty years (e.g., Hunter & Kearney, 1982; Flash & Hogan 1985; Mussa-Ivaldi, Hogan & Bizzi, 1985; Shadmehr & Mussa-Ivaldi, 1994; Burdet et al., 2001; Körding & Wolpert 2004; Pruszynski et al., 2011). Conversely, dynamic modeling of human motion has found application in the movie and gaming industries for realistic animation of characters. Also, collaborative robots are increasingly being developed to assist humans in their tasks and extend their physical capabilities (figure 1.1). These include exoskeletons that can amplify the forces exerted by physically impaired or normal individuals, virtual reality training systems for surgery or neurorehabilitation and brain–machine interfaces (BMIs). All of these systems physically collaborate with humans and require a good understanding of how humans control their limbs when interacting with the environment in order to be used effectively. The control of such systems demands tools that go beyond the descriptive models typically

Figure 1.1
Examples of robots mechanically interacting with humans. (A) Robotic device to help workers manipulate heavy and large objects, developed by Stanley Cobotics and Northwestern University (with permission from J. E. Colgate). (B) System for locked-in individual using EEG signal to control a wheelchair (adapted from Rebsamen et al. [2010] with permission). (C) Current training of microsurgery on rats (left) may be complemented by systematic training using virtual reality workstations (right) (reproduced from Wang et al. [2004] with permission). (D) A platform for robot-assisted post-stroke therapy using virtual reality therapeutic games to promote motivation and training (Lambercy et al., 2011).

developed in neurophysiology; it requires efficient computational algorithms that take into account human biomechanics and neural control.

There are already excellent books that describe human motor control and learning from the point of view of neural computation such as (Shadmehr & Wise, 2005). However, developing computational algorithms that could lead to advances in rehabilitation protocols and robot control algorithms requires consideration of more than neural computation. Modeling only the neural processes involved in motor control can lead to misinterpretations. For example, an incomplete understanding of the ability to modify the mechanical properties of the muscles or limbs could lead to incorrect attribution of changes in motor behavior to plasticity in neural circuits. On the other hand, description of the biomechanics alone cannot explain motor control. For example, it cannot adequately describe the robustness of dynamic balance during walking or adaptation to changes in the environment, such as compensation for windy conditions when kicking a ball. To understand the control of such tasks, we must consider both the sensory feedback and the modification of the control processes, just as in robotics and control theory. Therefore, in order to understand the dynamics of manipulation, it is necessary to interpret the neural processes with a balanced consideration of biomechanics and control.

The human arm, like a (humanoid) robot, has segments connected by joints, muscles as actuators, position and force sensors, and a neural control system (figure 1.2). A defining characteristic of robots is that of an integrated *system*; namely, robotics is the integration of mechanical and electrical hardware with control software and algorithms. As an analogy, we introduce the term *human robotics* to capture the idea of *a system level synthesis of biomechanics and neural control*. We do not view the neural system and limbs in isolation from each other; rather, we recognize that they form an integrated unit and must be studied in that context. Our approach is to develop relatively simple computational models that capture enough of the complexity of the biological system to serve as useful tools for investigating how the brain functions in learning to control movement, improving the control of neural prostheses, and enhancing the effectiveness of rehabilitation therapy. Furthermore, this approach may enable the development of future *cyborgs*—machines collaborating efficiently with animal systems and extending their capabilities.

Our approach to the study of motor control is illustrated in figure 1.2. Using a virtual reality workstation, psychophysical experiments can be carried out to compare goal-directed movements before and after altering visual feedback or environmental mechanics. For example, the robotic interface of figure 1.2A amplifies deviations of planar reaching arm movements from the straight line joining start to target, and we can observe the adaptation of movement to this load. In general, a typical approach in investigating motor control involves perturbing the movement (output) and/or sensory feedback (input) and analyzing the resulting modifications in the motor behavior. The effects of perturbations on limb trajectories, end-effector forces, and muscle activation patterns are then used to characterize the adaptation process. In many cases, interpretation of the data requires

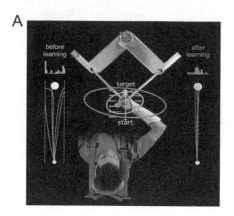

A

B

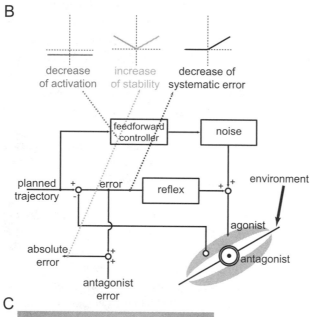

decrease of activation

increase of stability

decrease of systematic error

feedforward controller

noise

planned trajectory

+ −

error

reflex

environment

agonist

antagonist

absolute error

antagonist error

C

Figure 1.2
Observation of how humans learn to perform successful movements in novel environments created with the help of a robotic interface. (A) Amplification of the lateral deviation from a straight line to the target. Computational modeling of this learning (B) led to a novel algorithm for adaptive robot interaction with the environment (C). Reproduced with permission from Burdet et al. (2013).

development of a computational model that complements the signal processing and statistical analysis and can generate predictions to be tested in further experiments. For instance, we have developed a computer model of neural control and adaptation (figure 1.2B) to simulate and understand the results of figure 1.2A, as described in chapter 8. Computational models and control algorithms derived from motor control studies can in turn be used to design better collaborative robots (figure 1.2C), assistive devices, and training systems for surgery and robot-assisted neurorehabilitation.

This book analyzes the control of arm movements, in particular reaching movements that are fundamental to manipulation and activities of daily living. It does not cover topics related to the control of the lower limbs, walking, or balance. The models and algorithms of motor control it describes should enable the reader to simulate motor control and learning at the levels of the task and of the musculoskeletal control. On the other hand, it does not consider models of the neural mechanisms that implement this control or processes involved in visual control of movement. There are several good reference textbooks (Kandel, Schwarz & Jessell, 2000; Bear, Connors & Paradiso, 2007; Purves et al., 2012) that complement the neurophysiological material. The kinematics and dynamics in the three-dimensional space—which are not addressed in this book, as most of the experimental evidence available concerns planar arm movements—are treated in standard robotics books (e.g., Craig, 2005; Spong, Hutchinson & Vidyasagar, 2006; Siciliano et al., 2009). A simple treatment of nonlinear and adaptive control can be found in Slotine & Li (1991). Finally, Menzel & d'Alusio (2000) and Ichbiah (2005) show recent developments in robotic systems and humanoids.

1.2 Outline: How Do We Learn to Control Motion?

A major issue in describing motor control is the problem of integrating the one-way input–output processing involved in neural information for motion planning and execution with the two-way energy exchange that characterizes interaction of the arm with the environment (Hogan 1990) and the delayed neural feedback (figure 1.3). This book provides a comprehensive description of human motor control integrating these aspects, beginning with muscle mechanics and control and progressing in a logical manner to planning and behavior.

Let us consider the task of learning to play tennis. Initially, you have difficulty even hitting the ball with the racquet, but gradually, with practice, you learn to control the motion of the racquet and your ability to control the trajectory of the ball improves. What has changed that makes you a better player? You must first learn to control your muscles in order to move your body toward a desired goal. Your nervous system, whose functions will be described in chapter 2, must learn which muscles to activate, when to activate them, and how much activation is required. Learning requires sensation of the motion provided by visual, kinesthetic, proprioceptive, and tactile sensory signals (chapter 2). However,

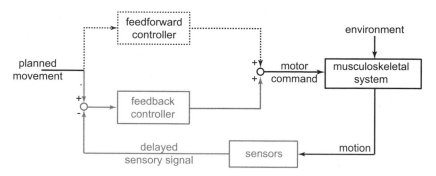

Figure 1.3
Schematic of human motor control. Two pathways can be seen: the feedforward control pathway (dotted lines) and the feedback control pathway (gray lines). Elements that form part of both pathways are shown in solid black lines.

any changes in the neural commands to muscles must consider the muscle mechanics (chapter 3) and feedback loops that interact with the "descending" neural commands (chapter 4).

To perform suitable movements, the motor control system must take into account the kinematics of the skeleton and the muscles that move it (chapter 5), as well as the muscle dynamics and dynamics of the moving body segments (chapter 6). It must also solve the problems of redundant degrees of freedom for both the articulated skeleton and the muscles that move it (chapters 5 and 6). To do so, the motor system is organized as a hierarchy of control levels, each provided with the sensory information for the functions that it controls, as described in chapter 2.

Furthermore, your central nervous system (CNS) must learn the dynamics of the ball (velocity, spin, rebound) and the mechanical properties of the racquet (center of mass, moment of inertia, stiffness). It must be able to judge the amount of force and timing of arm and leg movements necessary to contact the ball, which requires integration of visual input to predict ball movement. In addition, it must learn how to react to disturbances such as a change from the expected ball trajectory due to a gust of wind or misjudged ball spin. How humans learn compensation and prediction for novel dynamics is described in chapters 7 and 8.

Planning the racquet's movement is not trivial, as such planning must consider the current position while integrating sensory information about the ongoing action. Reactive motion planning is discussed in chapter 9. To predict the ball's movement, the CNS integrates visual information about ball movement coming from the eyes and information about arm and body position coming from muscles, skin, and joints in order to adjust commands, which are sent to muscles via motoneurons in the spinal cord. Chapter 10 describes how the CNS can combine different sensory signals to predict movement. Fundamentally, to control complex movements such those as needed to play tennis, the CNS uses both *feedback and feedforward control mechanisms* (figure 1.3). In *feedback control*, sensory information is used to control a process. In some cases, this process may involve a discrete reaction, such as the reflex withdrawal of the hand after unexpectedly touching a hot object. In other cases, it may involve intermittent correction to motor commands during an ongoing action that are modified according to new sensory input, such as steering a car to avoid hazards. In *feedforward control*, descending motor commands are generated according to a plan of the desired kinematics and the expected dynamics. Note that feedforward control can use feedback to modify the plan during its execution; that is, the plan can be contingent on the realized motion. Feedforward control is required because of feedback delays. Sensory information is not immediately available to the motor control centers because of time taken to generate and transmit sensory signals. In fast movements, there is insufficient time to process sensory information. For example, the reaction time to visual feedback is on the order of 150 milliseconds (ms), which may be as long as the time taken to execute a rapid limb movement.

1.3 Experimental Tools

Quantitative models of motor control heavily rely on experimental data provided by tools that will be presented later in this section, for both calibration and validation. Measurement of limb trajectories, forces exerted on the environment, muscle activation, and neural control require access to precise and systematic experimental techniques and devices.

Motion Capture

Soon after photography was developed, it became apparent that it could be used to capture the evolution of limb kinematics during human movements. Figure 1.4A shows an example of stroboscopic photography of walking taken by Etienne Marey near the end of the nineteenth century. This technique has evolved into sophisticated *camera-based motion capture systems* (figure 1.4B). In these systems, reflective (passive) or light-emitting (active) markers placed on the limb segments are recognized by the cameras in order to determine the position of the marker in Cartesian space. From these measurements, the software can reconstruct the motion of the limbs (kinematics), which can further be used to compute the dynamics using anthropometric models of body segment inertia and measurements of environmental reaction forces. Such motion capture systems are used not only for neuroscience studies but also in the entertainment industry. Although dynamic models of objects and animals have improved dramatically in recent years, in many cases it is still simpler and cheaper to record the movements of real actors in order to animate synthetic figures with motions that appear natural.

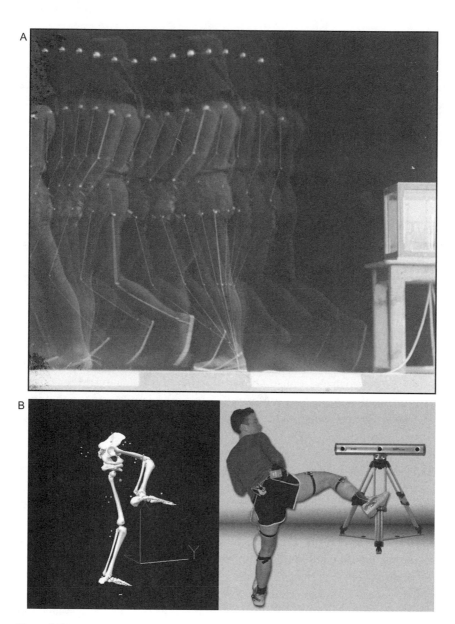

Figure 1.4
Motion capture. (A) Walking recorded by Marey circa 1880 (reproduced with permission from College de France). Clothes emblazoned with a white band along the limb and stroboscopic photography enabled visualization of limb and body movements during walking. (B) Measurement of whole body movement using a modern camera-based system (with permission from Northern Digital).

Inertial Measurement Units

Inexpensive and compact sensor units typically composed of an accelerometer, gyroscope, and compass have recently appeared on the market; these can be affixed to the body or integrated into clothing. Although computation of motion using measurements made with accelerometers typically suffers from the integrated error during a movement, combining information from a set of sensors fixed to the arm or to the body enables fairly accurate description of the movement. Furthermore, the information from such sensors can be combined with information from other sensors using Kalman filtering techniques (described in chapter 10).

Force

Reaction forces during interaction with the environment are often measured using six-axis force sensors, which can measure the force along three orthogonal axes as well as the torque around each axis. During activities such as walking, running, and jumping, ground reaction forces can be measured using force plates—rectangular plates with sensors sensitive to pressure or force mounted under each corner. Balance boards such as the Wii system marketed by Nintendo represent a rudimentary facsimile of a force plate but are noisier, less sensitive, and less accurate. Intelligent objects equipped with force sensors (and possibly inertial measurement units) can be used to study how humans manipulate these objects.

Electromyography (EMG)

We have known since Galvani's experiments in the late eighteenth century that motor commands to muscles are transmitted through changes of electrical potential. An indirect measure of the motor command to a muscle can be obtained by recording the potential difference between two electrodes placed on the skin overlying the muscle. Such surface electrode recording of electrical activity in muscles, known as *surface electromyography* (EMG), provides an estimate of the level of activation in a region of the muscle near the electrodes, which can be used to infer how muscle activity relates to movement (figure 1.5). A useful aspect of EMG is that the signal precedes the limb movement by several hundredths of a second because of the delay due to signal conduction and muscle mechanics. This feature can be exploited to trigger disturbances at or before movement onset.

Because of the stochastic nature of the process involved in the activation of muscle fibers, EMG signals are highly variable. The relative times at which different muscle fibers are activated vary from one movement to another even when the same movement is repeated. Thus, the EMG can be quite different for movements with very similar kinematics. For offline applications, the mean over a number of trials can be used to improve reliability. Rather than using surface electrodes, needles or fine wires can be inserted directly into the muscle. This approach reduces the possibility of contamination by electrical signals originating from neighboring muscles (called *cross-talk*) by improving

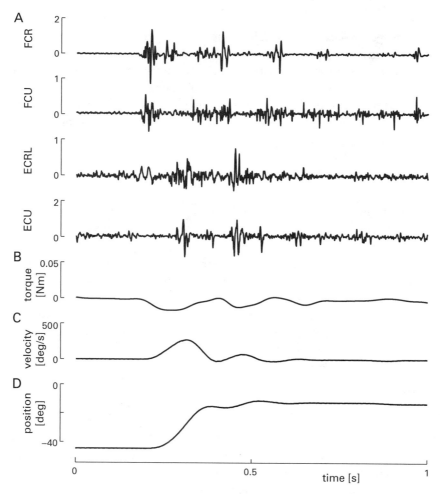

Figure 1.5
Human wrist flexion movements performed in the presence of a negative viscous load produced by a robotic interface (Milner & Cloutier, 1993). (A) EMG signals of flexor carpi radialis (FCR), flexor carpi ulnaris (FCU), extensor carpi radialis longus (ECRL), and extensor carpi ulnaris (ECU). (B) Torque between the robot and wrist, (C) angular velocity, and (D) angular position recorded by the interface's sensors.

localization of the signal compared to surface electrodes. The smaller the electrode surface area, the more selective the signal will be; that is, the smaller the region contributing to the signal. However, extreme selectivity can result in a signal that may not be representative of overall muscle activity, particularly in the case of large muscles.

Virtual Reality Workstation

Over the last twenty years, *virtual reality workstations* have been used increasingly often to investigate human motor control. Such workstations typically include a robot to which the hand or the arm is coupled in some manner (figures 1.1D, 1.2A), a visual display in the form of a computer-generated image either displayed on a monitor or projected on a surface, and auditory information often used for cuing responses or providing knowledge of results. Such workstations can create virtual dynamic, visual, and audio environments that simulate natural environments or produce unnatural conditions. For example, one can create environments in which visual feedback is incongruent with proprioceptive or audio feedback in order to examine how the brain combines information from various sensory channels. One can also create novel environmental dynamics to which the human must adapt in order to succeed in a task (figure 1.2A). Besides creating environments that are unnatural or incongruent, one of the features of these systems is the ability to switch the environment almost instantaneously. This feature allows the experimenter to introduce perturbations or employ other techniques that can measure features of the control processes or probe for changes in the mechanical properties of the system.

Functional Imaging

Various techniques have been developed to examine how brain activity is localized and modulated during different actions. One of these, *functional magnetic resonance imaging* (fMRI), is based on the metabolic requirement for neural activity, which creates a demand for oxygen. The vascular system responds by increasing the amount of oxygenated hemoglobin relative to deoxygenated hemoglobin. Deoxygenated hemoglobin has paramagnetic properties that create nuclear magnetic resonance when excited by appropriate electromagnetic perturbations. The strength of the resonance signal depends on the concentration of deoxygenated hemoglobin. A large background magnetic field induced by a scanner aligns the magnetic spin axis of the deoxygenated hemoglobin molecules, which is then perturbed by an electromagnetic pulse. The relative strength of this signal in different brain areas is presumed to be a representative measure of their relative neural activity.

A recent approach to investigate the neural correlates of motor learning involves training with a robot in the laboratory followed by resting-state fMRI in a scanner located elsewhere. Persistent change in the correlation of spontaneous activity between brain regions in the resting state implies changes in the functional connectivity between those regions. By quantifying the change in the correlation (the strength of functional connectivity), it is possible to infer which brain regions are most involved in the process under investigation.

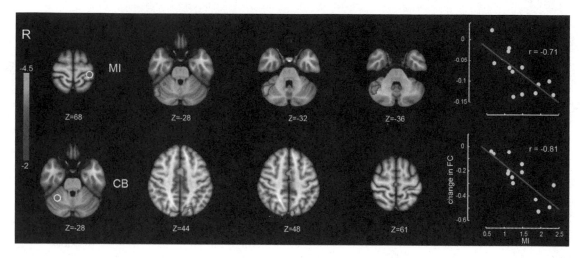

Figure 1.6
Resting-state activity recorded using fMRI of the whole brain can be used to analyze the connectivity between different regions of the brain. Following motor learning, functional connectivity between regions of the senso-rimotor network increases in strength. The top row shows that the change in functional connectivity (FC) between the primary motor cortex (white circle) and cerebellum (shaded regions in subsequent three plots) is positively correlated with the amount of motor learning (MI), as shown on the right. In contrast, the bottom row shows that there is a negative correlation between the change in FC between the cerebellum (white circle) and motor regions, as well as the superior parietal lobule (shaded regions in subsequent three plots). Reproduced with permission from Vahdat et al. (2011).

For example, figure 1.6 shows how the change in functional connectivity between the primary motor cortex and cerebellum is highly correlated with the amount of adaptation to a change in environmental dynamics suggesting that the information being passed between these two regions is principally responsible for motor learning.

Electroencephalography (EEG)
Electroencephalography (EEG) measures differences in the electrical potential between two locations on the scalp, which are produced by the changing electrical activity of neurons in the brain. EEG can be used to detect different states of the brain by analyzing the amplitude of the recorded signal in different frequency bands. Portable systems can be used as interfaces to control a computer cursor or a robot such as the wheelchair shown in figure 1.1B. Such systems can be used to facilitate communication and mobility for persons with extreme weakness or paralysis in conditions such as amyotrophic lateral sclerosis (ALS) or tetraplegia. EEG has excellent temporal resolution with negligible delay, but its spatial resolution is relatively poor; that is, the source of the signal cannot be accurately determined and is limited to the cortical surface of the brain. In contrast, the spatial resolution of fMRI is in the order of millimeters throughout the entire brain, but its temporal resolution is low with a relatively long delay (on the order of a few seconds).

Because EEG and fMRI have complementary properties, it is sometimes advantageous to combine them.

Transcranial Magnetic Stimulation (TMS)

Transcranial magnetic stimulation (TMS) is a noninvasive technique that can be used to stimulate the neurons of the central or peripheral nervous systems. TMS produces a brief magnetic pulse in a coil directly above the area of the nervous system to be stimulated. This pulse induces an electrical current in the neurons, causing them to fire. When this process is performed over the motor cortex of the brain, it can produce muscular twitches. Depending on the stimulation parameters, the application of TMS can produce either excitation or inhibition of the nearby neural circuits. Because TMS can produce excitation or even disrupt the processing of particular areas of the cortex, it can be used to study cortical functions.

1.4   Summary

In this chapter, we have explained the motivation for a transdisciplinary approach of (musculoskeletal) biomechanics and (neural) control to investigate human motor control. We believe that human movement dynamics can be modeled using computational algorithms of motor control that incorporate the fundamental mechanical features of the musculoskeletal system, the transduction characteristics of sensory receptors, and critical elements of the neural control circuitry. We briefly described the concepts, tools, and techniques used to develop this human robotics approach. This book focuses on the insights that this approach has provided in understanding how movement of the arm is controlled and how the control adapts to changing environments. The simulations that have been used to generate some of the figures of this book are available on Etienne Burdet's website and can be used to simulate the control of the human arm under a variety of conditions.

2 Neural Control of Movement

The objective of this chapter is to provide a brief overview of the architecture and function of the nervous system as it relates to sensing and control. The terminology is that of neuroscience, as the intention is to provide a framework and language that will allow roboticists and neuroscientists to communicate effectively. There is much less detail, particularly in the description of molecular processes, than would normally be found in a neuroscience textbook. However, the fundamental concepts necessary for developing a meaningful model of human motor control are presented. Detailed models of computational processes that may be implemented in specific regions of the brain are beyond the scope of this textbook. The reader is referred to recent work such as that of Dean and Porrill (2011) on computational modeling of the cerebellum, which provides considerable insight into the detailed information available for modeling, or the review of Gao, van Beugen, and De Zeeuw (2012) on synaptic plasticity, which examines the complex circuitry available for processing neural signals in the context of cerebellar learning.

2.1 Bioelectric Signal Transmission in the Nervous System

In robotic systems, sensors may represent information in either an analog or digital form, although information is generally processed in a digital form by the controller. Similarly, the commands issued by the controller are generally digital values that are converted to analog outputs to the actuators. Although analog controllers were the norm at the advent of robotics, today controllers generally perform computations using digital information. The human nervous system also transmits and processes signals that are effectively digital in nature. In robotics, the digital signals are generally the binary representation of an analog value. These are transmitted along parallel lines as their decomposed sum of powers of 2, that is, 2^0, 2^1, 2^2, and so on, multiplied by a coefficient (0 or 1). In the nervous system, the information is transmitted along a serial line in which the value is encoded in terms of the frequency of electrical impulses called *action potentials*.

Action potentials are generated and transmitted by specialized cells called *neurons* (Kandel, Schwarz & Jessell, 2000). The signal transmission path for voluntary activation of muscles originates in specific regions of the brain and follows a route that passes through the brainstem to the spinal cord and then on to muscles. The brain, brainstem, and spinal cord constitute the CNS and can be considered as a hierarchical controller in which neurons in the brain can modulate the output of neurons in the brainstem and spinal cord, whereas neurons in the brainstem principally modulate the output of spinal cord neurons. Together, the neurons in the brain, brainstem, and spinal cord constitute the CNS. Neurons throughout the CNS receive and integrate signals originating from sensory neurons, usually referred to as sensory receptors, located in various body tissues.

A neuron is bounded by a lipid membrane that separates the intracellular and extracellular fluids. The membrane is essentially impermeable to electrically charged molecules such as ions. The concentration of potassium ions [K^+] is markedly higher and the concentration of sodium [Na^+] ions is markedly lower in the intracellular fluid than the extracellular fluid. The force created by this ionic concentration difference is balanced by an electrical potential difference across the membrane. In the steady state, the potential difference across the membrane of the neuron is about −70 mV to −80 mV. This steady-state potential difference is referred to as the *resting membrane potential*.

The electrically charged ion attracts a cloud of water molecules that effectively prevent it from passing across the hydrophobic lipid membrane barrier that separates the extracellular and intracellular fluid. The only way that ions can move across the cell membrane is by facilitated diffusion involving specialized protein molecules embedded in the membrane. Some of these protein molecules are organized as complexes of subunits, which can form pores or *ion channels*. The specific structure of a protein molecule—that is, the sequence of amino acids that make up the peptide chain of the molecule—determines its function. In the case of an ion channel, the amino acid sequences of the various subunits create regions with strong electric fields that allow the channel to be selective for specific ions, such as K^+ or Na^+.

An action potential represents a rapid change in the difference in electrical potential between the intracellular and extracellular fluid (figure 2.1). Ion channels are normally in a closed state and can be opened only by a specific triggering event. The channel opening is triggered when the membrane potential becomes sufficiently less negative than the resting potential (*depolarization*). Ion channels that open in response to a change in membrane potential are called *voltage-gated channels*. Opening and closing of ion channels is a random process whose probability is a function of the membrane potential. When the membrane potential reaches a threshold of about −55 mV, there is a high probability that the protein subunits of voltage-gated Na^+ channels will change their conformation, creating an open state that allows rapid diffusion of Na^+ from the extracellular to the intracellular fluid. This diffusion transiently depolarizes the membrane to about +30 mV. The Na^+ channels close a fraction of a millisecond after opening, at which point K^+ channels open, causing the membrane potential to drop below its resting value as K^+ diffuses from the intracellular to the extracellular fluid. During this time, specialized proteins known as Na^+/

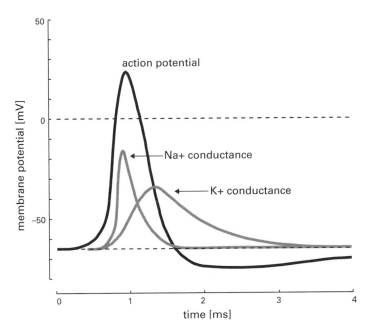

Figure 2.1
The depolarization phase of the action potential is initiated when Na⁺ channels open and Na⁺ conductance transiently increases. This phase is followed shortly by opening of K⁺ channels and a long-lasting increase in K⁺ channel conductance that repolarizes the membrane and leads to a period of afterhyperpolarization during which the membrane potential is more negative than the resting membrane potential.

K^+ pumps remove Na^+ from the intracellular fluid and return K^+. After several milliseconds, the K^+ channels close and the resting membrane potential is restored. The interval during which the membrane potential is more positive than its resting value lasts 1–2 ms and is followed by an interval of about 2 ms during which the membrane potential is below its normal resting value. Until the membrane potential is restored to its resting value, the probability of a second action potential occurring is very low because of the low probability of ion channel reopening. The period of low probability is called the *refractory state*. The reciprocal of the interval between successive action potentials is referred to as the instantaneous *firing rate* of the neuron.

A neuron can be separated into three primary structural regions (figure 2.2): the *dendrites*, which receive, integrate, and process input from other neurons; the *soma* or cell body, which can also receive signals and further process the integrated information originating from the dendrites and which is the site of genetic and metabolic activity; and the *axon*, which transmits an output signal to targets. In the case of a *motoneuron*, the axon acts purely as a transmission line to its target, the muscle fibers. As long as there is no conduction failure along the axon, the output signal generated at the soma is transmitted

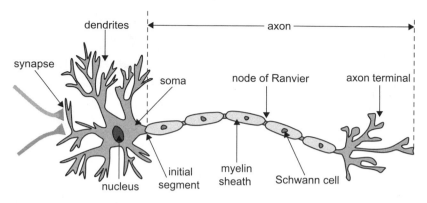

Figure 2.2
Schematic representation of a neuron illustrating the three principal compartments: dendrites, soma, and axon (adapted from http://en.wikipedia.org/wiki/Neuron).

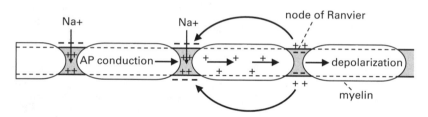

Figure 2.3
An action potential propagates along an axon as depolarization spreads along the axon membrane. This process occurs very rapidly in myelinated regions where the change in membrane potential is primarily capacitive and more slowly at nodes of Ranvier, where capacitance is lower and depolarization occurs primarily by opening of voltage-gated Na^+ channels that effectively regenerate the depolarization at each node.

faithfully to the target muscle fibers. However, the output of neurons that target other neurons may be attenuated or preempted prior to reaching the target neuron. This process, *presynaptic inhibition*, is described later in this chapter.

Ionic and capacitive currents generated by the electric field associated with an action potential cause membrane depolarization to conduct from the site of Na^+ channel opening. Although the depolarization decrements exponentially with distance, adjacent regions of the membrane reach the threshold required for Na^+ channel opening. In this way, an action potential can propagate along the membrane from the site of initiation at the soma to the *axon terminals* (figure 2.3). The speed at which the propagation occurs is regulated by the membrane length and time constants, which respectively determine how much adjacent membrane segments are depolarized and how quickly the depolarization occurs. The largest nerve fibers have the highest propagation speeds (*conduction velocities*), not only because of their diameter but also because they are surrounded by a myelin sheath consisting of tightly wrapped layers of insulating myelin, which facilitates rapid passive spread of depolarization due to its low capacitance. However, if this sheath were continuous from the soma, the depolarization would be negligible by the time it reached the axon terminal. Therefore, the sheaths are interrupted at regular intervals by patches of bare membrane (*nodes of Ranvier*) that contain large numbers of Na^+ and K^+ channels, allowing the action potential to be intermittently regenerated along the axon until it arrives at the terminal.

2.2   Information Processing in the Nervous System

Signal transmission from one neuron to another or from a motoneuron to a skeletal muscle fiber is achieved by diffusion of molecules released from an axon terminal to receptor molecules embedded in the membrane of the target cell. When a neurotransmitter molecule binds to a receptor, it causes an ion channel to open, resulting in a change in membrane potential. The specialized region where this process occurs is called a *synapse*. The membrane of the axon terminal is called the *presynaptic membrane*; the membrane where the receptors are located is called the *postsynaptic membrane*. The entry of Ca^{2+} into the axon terminal through voltage-gated ion channels is responsible for triggering the release of neurotransmitter molecules from storage vesicles. Opening of ion channels associated with receptors in the postsynaptic membrane will depolarize the postsynaptic membrane if this process results in a net influx of positive ions, such as Na^+ or Ca^{2+}, which are both more highly concentrated in the extracellular than the intracellular fluid. In contrast, the postsynaptic membrane will *hyperpolarize* (the membrane potential will become more negative) if this process results in an net efflux of positive ions such as K^+, which is more highly concentrated in the intracellular than the extracellular fluid, or a net influx of negative ions such as Cl^-, which is more highly concentrated in the extracellular than the intracellular fluid. Depolarization of the postsynaptic membrane is referred to as *excitation* and hyperpolarization is referred to as *inhibition*.

To explain how muscles are controlled, we will use a bottom-up approach tracing the control signal from a muscle fiber back to the brain. Muscles consist of bundles of individual fibers whose mechanical properties are described in considerable detail in chapter 3. Groups of tens or hundreds of muscle fibers, distributed relatively randomly throughout the muscle, are controlled by a single neuron located in the spinal cord called an α-*motoneuron*. A group of muscle fibers and the α-motoneuron, which controls them form a *motor unit* (Heckman & Enoka, 2012). The soma and dendrites of motoneurons are located entirely within the spinal cord; the axons leave the spinal cord in bundles called nerves. Individual axons are often referred to as *nerve fibers*. Groups of nerve fibers split from the main nerve trunk at various locations in the body and follow specific routes to the muscles that they control. Each axon branches as it enters the muscle, with each branch forming a synapse (called a *neuromuscular junction*) with one muscle fiber. Each muscle fiber is controlled by a single axon, that is, by a single α-motoneuron.

Motoneurons, as well as other neurons in the spinal cord, brainstem, and brain receive feedback from *sensory receptors* located in the tissues. These sensory receptors are specialized dendrites at the end of long nerve fibers. Sensory receptors located in the skin, muscle, and connective tissue transform mechanical events into action potentials. The somas of these sensory neurons are located just outside of the spinal cord and send axons into the spinal cord. Together, the nerve fibers from sensory receptors and motoneurons that are located outside of the spinal cord form the *peripheral nervous system*.

The α-motoneurons receive inputs from a multitude of other neurons, including peripheral sensory neurons—some of which are excitatory and some of which are inhibitory. The integration of these inputs determines the motoneuron output, that is, the motor unit control signal. The control signal is a train of action potentials in which the amount of motor unit force that it specifies is inversely related to the interval between successive action potentials; that is, the shorter the interval (the higher the firing rate), the greater the force produced by the muscle fibers of the motor unit. The mechanism by which the electrical signals (action potentials) are converted to force is described in chapter 3.

To understand how an α-motoneuron integrates its diverse inputs, consider the change in membrane potential at a single synapse. Although thousands of neurotransmitter molecules are released from the presynaptic axon terminal, not all of them bind to receptors. The net effect of the ion channels that they open is a small change in membrane potential that, by itself, cannot significantly alter the firing rate of the postsynaptic neuron. Modification of the postsynaptic firing rate requires the combined effect of many synapses. The overall change in membrane potential of the postsynaptic membrane is determined by the integration of nearby synaptic input received over a short time interval, that is, the *spatial and temporal integration* of synaptic input that depends on the membrane length and time constant, respectively. Depending on the relative amount of excitatory and inhibitory synaptic input to the postsynaptic neuron, the overall effect may be an increase (net excitation) or decrease (net inhibition) in firing rate. In the case of an α-motoneuron, any change in firing rate represents a change in the force command to the muscle fibers that it controls.

An important characteristic of synaptic integration is that it is characterized by *stochastic processes* and is hence inherently variable—that is, noisy. Several factors contribute to synaptic noise. The concentration of neurotransmitter in the synaptic space following a presynaptic action potential will vary depending on the amount of neurotransmitter released spontaneously prior to the action potential, the amount released by the action potential, the amount that leaves the synaptic space due to molecular diffusion, the amount of reuptake of neurotransmitter by the presynaptic membrane, and the concentration of enzymes in the synaptic cleft that degrade the neurotransmitter. This combination of factors introduces variability in the amount of neurotransmitter that binds to receptors in the postsynaptic membrane. Another important factor contributing to variability is the stochastic nature of the opening of voltage-gated ion channels, where the probability is a function of the membrane potential. Voltage-gated channels are found in both the presynaptic and postsynaptic membranes and are a secondary feature of NMDA-gated channels that require membrane depolarization to remove a Mg^{2+} ion that blocks a cation channel at resting membrane potential. Thus, the concentration of receptors on a patch of postsynaptic membrane that will be activated during a presynaptic action potential will vary, leading to variation in the size of postsynaptic potentials in different regions of the dendrites. Spatial and temporal summation of the postsynaptic potentials will tend to reduce the variability by integration. However, the process of triggering an action potential in the postsynaptic neuron will again contribute to variability because the opening of voltage-gated Na^+ channels at the site of action potential generation (initial segment) is determined by a probability function of the membrane potential.

2.3 Peripheral Sensory Receptors

A considerable amount of the synaptic input to α-motoneurons originates from peripheral sensory receptors located in the skin, muscles, and joints. Most of these receptors are sensitive to mechanical stimuli such as strain and pressure and are therefore referred to as *mechanoreceptors*. Mechanoreceptors are specialized sensory nerve fiber terminals that provide the CNS with tactile, proprioceptive, and kinesthetic information. They encode stimuli by means of changes in membrane potential produced by opening of mechanically gated ion channels located in the nerve fiber terminal. These ion channels open as the result of mechanical deformation of the nerve fiber terminal. Movement of ions through the channels leads to membrane depolarization known as a *receptor potential*. If the receptor potential is large enough—that is, if the stimulus is sufficiently strong—the membrane depolarization will spread far enough to trigger the opening of voltage-gated Na^+ channels and the generation of an action potential that is conducted from the sensory receptor along the nerve fiber to the CNS through the spinal cord. The number of channels that open and hence the level of depolarization increases with the amount of mechanical deformation, that is, with the strength of the stimulus. The firing rate of the sensory receptor is roughly proportional to the magnitude of the depolarization.

Cutaneous Mechanoreceptors

Tactile sensation originates from mechanoreceptors in the skin that are sensitive to pressure, strain, and vibration (Iwamura, 2009). There are four principal types of receptors, which differ in sensitivity due to their physical structure and their location (figure 2.4). The structure—particularly the way in which the sensory nerve fibers are encapsulated—determines the types of stimuli that are most effective in mechanically deforming the nerve

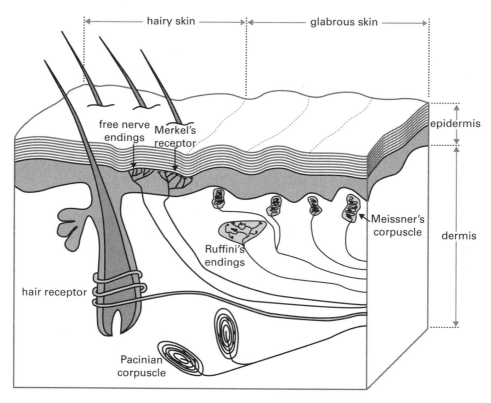

Figure 2.4
Cutaneous receptors are found in superficial layers of the skin (Merkel's receptors, Meissner's corpuscles and free nerve endings) and in deeper layers (Pacinian corpuscles and Ruffini's endings). Their location and structure determine their response to mechanical stimuli such as pressure and vibration, which leads to opening of ion channels through mechanically induced changes in channel conformation (adapted from http://en.wikibooks.org/wiki/Sensory_Systems/Somatosensory_System).

fiber terminals. It also determines the size of the receptive field, that is, the region of the skin over which a stimulus can activate the nerve fiber.

Merkel's receptors and *Meissner's corpuscles* are located near the boundary of the epidermal and dermal layers of the skin. Merkel's receptors are sensitive to light pressure, which signals the contact of an object with the skin. They have small receptive fields, so variation in pressure sensed by a population of Merkel's receptors can provide information about object shape, particularly in detection and discrimination of edges, corners, and curvature. Meissner's corpuscles are sensitive to small deformations of the skin but have much larger receptive fields than Merkel's receptors. They are sensitive to low-frequency vibration in the 3–40 Hz band, which enables discrimination of rough surface texture and detection of object slip, critical for controlling grip force.

Pacinian corpuscles and *Ruffini's endings* are located deeper in the dermal layer of the skin. Pacinian corpuscles are even more sensitive to small indentation of the skin than Meissner's corpuscles. Consequently, they are sensitive to high-frequency vibration in the 250–350 Hz band, which enables discrimination of fine surface texture. However, because of their high sensitivity, they have very large receptive fields, so they provide limited information about the location of the stimulus. Ruffini's endings are sensitive to stretch of the skin that contributes to *proprioception*, that is, sense of joint position.

Sensory receptors are classified as rapidly or slowly adapting according to their response to a sustained stimulus (Johansson & Flanagan, 2009). A rapidly adapting receptor is sensitive to change but provides limited information about the magnitude of a sustained stimulus. Generally, these receptors transiently increase their firing rate at the onset and offset of the stimulus but are silent during the period that the stimulus is sustained (figure 2.5). A slowly adapting receptor also transiently increases its firing rate at the onset and offset of the stimulus, but the change in firing rate tends to be less than that of a rapidly adapting receptor. Furthermore, it fires at a relatively constant rate while the stimulus is sustained that is lower than the transient response but is roughly proportional to the magnitude of the stimulus, provided that the stimulus is not strong enough to produce saturation. Meissner's corpuscles and Pacinian corpuscles are rapidly adapting receptors, whereas Merkel's receptors and Ruffini's endings are slowly adapting receptors.

Muscle Mechanoreceptors

Muscle contains two important mechanoreceptors, *muscle spindles* and *Golgi tendon organs* (Gandevia, 2011; Prochazka, 2011; Proske & Gregory, 2002). Muscle spindles respond to stretch. They consist of small muscle fibers, referred to as *intrafusal* fibers, entwined by sensory nerve endings encapsulated together in spindle-shaped structures that are distributed throughout a muscle in parallel with skeletal muscle fibers, sometimes referred to as *extrafusal* fibers (figure 2.6). There are two types of sensory endings associated with intrafusal muscle fibers: primary endings (*group Ia*), which are large, fast-conducting nerve fibers; and secondary endings (*group II*), which are smaller, slower-conducting nerve fibers.

The mechanical properties of intrafusal fibers determine the proportion of an applied stretch that will act on the sensory endings and consequently how the stretch will be transduced into action potentials. The sensitivity of primary and secondary sensory endings appears to be directly related to local stretch in the intrafusal fibers and can be altered by changes in intrafusal fiber stiffness. The polar zones (regions near the ends) of an intrafusal fiber are considerably stiffer than the sensory zones, which are located near the center of the fiber. Therefore, when a muscle spindle is stretched, most of the change in length takes place in the sensory zone. Intrafusal fibers are usually innervated by *γ-motoneurons*. Stimulation of a γ-motoneuron produces localized contraction of an intrafusal fiber (usually

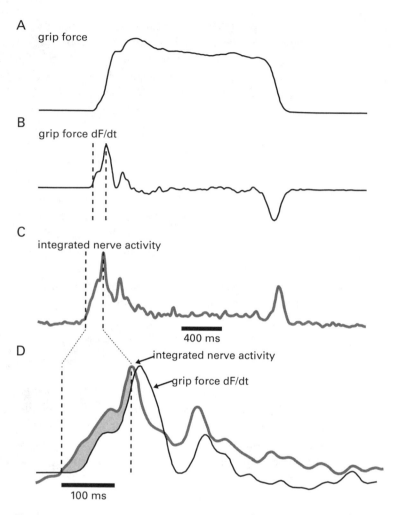

Figure 2.5
During transient changes in grip force (*A*), which occur at the onset and termination of the grip, the integrated
activity originating from cutaneous sensory receptors (*C*) parallels the absolute value of the rate of change of
grip force (*B*) but leads it by 10–30 ms, as can be seen when (*B*) and (*C*) are overlaid on an expanded time scale
(*D*). During periods of sustained grip force, the activity increases with the strength of the grip (adapted from
Milner et al. [1991]).

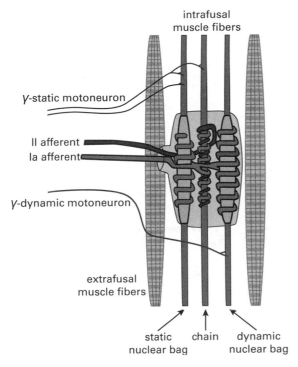

Figure 2.6
Muscle spindles can contain up to three different types of intrafusal muscle fibers that are innervated by different sensory nerve fibers and γ-motoneurons. The different mechanical properties of the intrafusal fibers, together with their γ-motoneuron innervation, endow the muscle spindle with a variety of responses to muscle stretch.

near the poles). The contraction stretches the sensory zone and also enhances its sensitivity to stretch because the polar zones become relatively stiffer and hence more resistant to stretch. Within the γ population, there are two distinct subpopulations: *static* and *dynamic*. The terms "static" and "dynamic" refer to the way in which stimulation of these γ-motoneurons influences the sensitivity of the muscle spindle.

During a ramp change in muscle length, both the primary and secondary endings show a discontinuous jump in firing rate at the onset of stretch (Prochazka, 2011). This phenomenon imparts a high sensitivity to the onset of muscle stretch; that is, the muscle spindles—particularly their primary endings—are excellent movement detectors. The firing rate of the primary ending drops abruptly after the initial transient and together with the secondary ending tracks the length change with a gain that is dependent on velocity (figure 2.7). The drop in firing rate of primary endings is likely due to a decrease in the stiffness of the polar zones of the intrafusal fibers, which results in a smaller proportion of the stretch being transmitted to the sensory zone. At the end of the stretch, while muscle length is being held constant, the firing rate of the secondary ending remains constant at

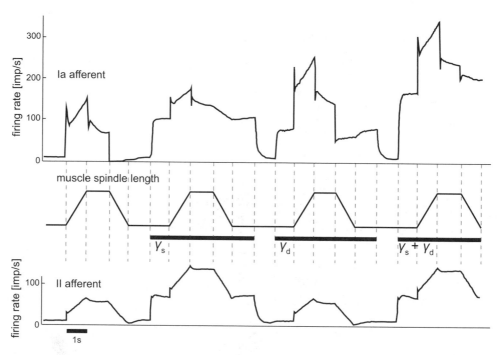

Figure 2.7
Muscle spindle afferent firing rates increase during muscle stretch and decrease during muscle shortening. Muscle spindle group Ia afferent fibers are particularly sensitive to the rate of change of length and show a large transient increase in firing rate at the onset of stretch. In the absence of γ-motoneuron activity, they tend to fall silent during muscle shortening. Muscle spindle group II afferent fibers tend to modulate their firing rate in proportion to muscle length (adapted from Prochazka [2011]).

its new level, whereas the firing rate of the primary ending drops abruptly and then declines more gradually to a new steady rate. This behavior highlights the velocity sensitivity of the primary endings. When the length begins to decrease, the primary ending falls silent while the firing rate of the secondary ending tracks the change in length. Although the average firing rate of primary endings increases with the velocity of stretch, the increase in firing rate is less than linear, increasing approximately as the cube root of velocity: $v^{1/3}$.

In the presence of γ-static motoneuron activation, both the primary and secondary endings increase their baseline firing rate. The sensitivity of the secondary ending to changes in length is somewhat enhanced, whereas that of the primary ending is considerably reduced. The secondary ending is unaffected by γ-dynamic motoneuron activation, whereas the primary ending increases both its baseline firing rate and its sensitivity to the constant velocity stretch (figure 2.7). One consequence of γ-motoneuron activation is that the primary endings do not fall silent during muscle shortening. If α-motoneurons and γ-motoneurons are coactivated, the CNS can receive information from primary endings during both muscle lengthening and muscle shortening.

Primary endings exhibit nonlinear amplitude dependent behavior, which is clearly evident in the response to a sinusoidal length change. As the amplitude of the stretch increases, the sensitivity of the sensory endings decreases. The sensitivity of primary endings within the small amplitude linear range may be from ten to a hundred times greater than that observed with larger stretches. After a large amplitude stretch, the primary ending can reset itself so that the high value of sensitivity reappears within several seconds after reaching the new length. This property is attributed to a change in the stiffness of the polar regions of the intrafusal muscle fibers. For small amplitude length changes, stiffness of the polar region is relatively high, thereby enhancing the proportion of the stretch that is transmitted to the sensory zone. However, if the stretch is sufficiently large, the stiffness is reduced, hence reducing the sensitivity of the sensory zone. This effect is somewhat analogous to static and dynamic friction. Once a new length is reached and maintained, the higher stiffness of the polar regions is reestablished and the sensitivity of the sensory zone is restored. Stimulation of γ-dynamic axons during sinusoidal stretching results in an increase in modulation of the firing rate of primary endings.

Golgi tendon organs are mechanoreceptors that respond to force. They consist of bundles of collagen strands and sensory nerve endings enclosed in connective tissue capsules located at junctions between skeletal muscle fibers and a *tendon* or *aponeurosis*. Tendon organs are located both at superficial muscle-tendon junctions and at deep intramuscular muscle-tendon junctions (figure 2.8). The collagen strands originate from one

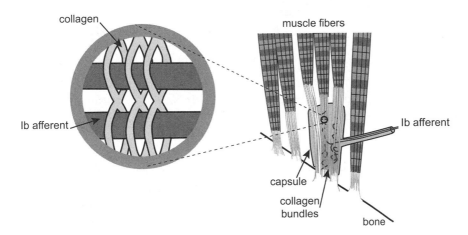

Figure 2.8
Golgi tendon organs are endings of Ib afferent nerve fibers that terminate among collagen fibers near the junction between muscle fibers and tendon. Increases in tension in the collagen fibers deform the nerve endings, producing depolarization, which leads to action potential initiation.

or more compact bundles that leave the tendon or aponeurosis. These bundles split further into compartments within the tendon organ capsule. Within each compartment, the collagen breaks up into fine strands that entwine the sensory nerve endings. At the muscle end, the strands rejoin to form tendinous attachments for muscle fibers. Generally, from ten to twenty-five muscle fibers attach in this way to form series connections with a single tendon organ. The muscle fibers attached to a tendon organ originate from several motor units. Muscle fibers from both slow- and fast-twitch motor units may connect to the same tendon organ.

Force is transmitted from the tendon and muscle ends of the Golgi tendon organ to its sensory endings through the collagen bundles and capsules in which they are embedded. The sensory nerve fibers that innervate the tendon organs are known as group *Ib afferents*. They begin to fire at the onset of contraction of in-series muscle fibers. If the in-series force applied to the Golgi tendon organ is sustained, afferent firing will be maintained. Golgi tendon organs have relatively low sensitivity to tension generated by passive stretching of the tendon but are highly sensitive to forces generated by active contraction of in-series muscle fibers.

Because the tendon is relatively stiff in comparison to the stiffness of passive muscle, the sensitivity to force produced by stretch of a passive muscle will be less than that produced by stretch of an active muscle. In a passive muscle, most of the stretch will be taken up by the muscle fibers rather than the tendon. Even in active muscle, though, sensitivity to tension produced by passively stretching the whole muscle will be less than that produced by contraction of in-series muscle fibers because most of the tension produced by passive stretching will be shunted through the in-parallel muscle fibers and hence will not be transmitted to the Golgi tendon organ.

Golgi tendon organs may be sensitive enough to fire one action potential in response to the twitch of a single in-series muscle fiber. A study of the response of Golgi tendon organs to motor unit contractions showed that under isometric conditions, they can respond preferentially to the dynamic component of this stimulus even in instances in which the static component appears quantitatively stronger (Jami, 1992). This dynamic sensitivity is apparently related to the speed of twitch contraction of the motor unit. Motor units whose rate of tension rise is low do not exhibit the 1:1 driving that characterizes dynamic sensitivity.

Individual Golgi tendon organs cannot signal whole muscle force under static conditions for several reasons. When several motor units influencing a single Golgi tendon organ are simultaneously activated, the resulting Ib firing rate is less than the sum of the firing rates when each motor unit is activated on its own, even though the force exerted on the Golgi tendon organ is the sum of the individual motor unit forces. In-parallel muscle fibers can produce unloading of the Golgi tendon organ and reduce the firing rate as whole muscle force increases. Therefore, the force sensed by an individual Golgi tendon organ does not represent a constant fraction of the whole muscle force. The mean firing rate of all Golgi tendon organs in a muscle may be less affected by unloading because a motor unit that unloads one Golgi tendon organ likely loads another. The mean firing rate tracks variations in motor unit force better than absolute levels of force; that is, Golgi tendon organs are more sensitive to dynamic changes in force than to static levels of force. Thus, they are better sensors of rate of change of force than force. Nevertheless, the sum of the combined firing rate of an ensemble of Golgi tendon organs appears to vary linearly with force at the tendon (Gandevia, 2011).

2.4   Functional Control of Movement by the Central Nervous System

The CNS consists of the spinal cord, brainstem and brain, each of which is composed of hundreds of millions to billions of neurons, many of which are involved in the control of movement. The output neurons that send control signals directly to muscles are primarily located in the spinal cord and are sometimes referred to as lower motoneurons. In addition to synaptic input from sensory receptors, they receive input from local circuit neurons in the spinal cord called *interneurons* and neurons in the brainstem and brain referred to as *upper motoneurons*. The ways in which sensory receptors and interneurons exert their influence on motoneurons are described in chapter 4. Interneurons, many of which are involved in coordinating the simultaneous activity of motoneurons of different muscles, also receive synaptic input from the brainstem and brain. The neural pathways from brainstem upper motoneurons to lower motoneurons are directed principally to muscles involved in maintaining balance (Horak, 2009). They act to compensate for disturbances to balance that originate from sources outside of the body or from the motion of body segments during purposeful actions. This compensation is achieved by the integration of sensory signals from the vestibular system, the visual system, and the peripheral nervous system, together with control signals originating from the cerebellum and the motor cortex. Brainstem upper motoneurons are involved both in preparatory activation of muscles in anticipation of disturbances to balance and in reflex circuits that attempt to restore balance following an unexpected or misjudged disturbance. The synaptic targets of these neural pathways are extensor muscles of the lower limbs, proximal muscles of the upper limbs, and trunk muscles.

Upper motoneurons in the brain exert control over both brainstem upper neurons and lower motoneurons. This control is achieved through network connections between regions that specialize in different functions. Our investigation of functional connections between brain regions involved in motor learning and sensory perception suggests that the network involved in motor control includes the *cerebellum, striatum (basal ganglia), primary somatosensory area, secondary somatosensory area, primary motor cortex, premotor cortex, supplementary motor area,* and *posterior parietal cortex* (Vahdat et al., 2011). However, these regions are not all functionally connected with each other. Rather, functional connections exist only between specific regions, reflecting the processing and

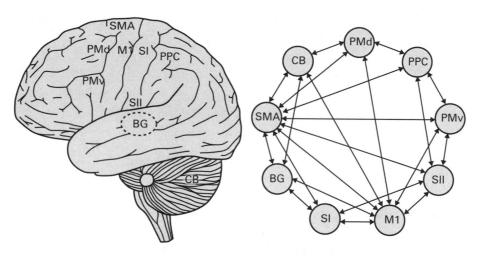

Figure 2.9
Network of connections known to be important in motor learning: dorsal premotor cortex (PMd), cerebellum (CB), supplementary motor area (SMA), basal ganglia (BG), primary somatosensory cortex (SI), primary motor cortex (M1), secondary somatosensory cortex (SII), ventral premotor cortex (PMv), and posterior parietal cortex (PPC). These connections show increases in the strength of functional connectivity after a motor learning task.

transfer of information required for planning and executing movements (figure 2.9). Furthermore, the strength of these connections is labile and can be adapted for different activities. A current hypothesis about how the CNS simplifies motor control suggests that the output of this network, emanating from the motor cortex, coordinates groups of muscles by engaging spinal interneurons (Bizzi et al., 2008). A network of spinal interneurons can act to regulate the synaptic input to each muscle in a group, governing its relative activation. When the relative activation pattern of a group of muscles is relatively invariant, the entity is referred to as a *synergy*. The idea is that the higher-level control can combine synergies in different ways by selecting the relative timing and strength of each synergy to achieve the large repertoire of movements required for different activities. The roles played by the different regions involved in higher-level motor control are described beginning with the cerebellum.

The cerebellum is a structure attached to the brainstem that receives input from many of the same sources as the brainstem. However, it does not integrate the information in the same way as the brainstem. Sensory signals originating from different sensory systems, such as vestibular, proprioceptive, and visual, are directed to distinct regions of the cerebellum and appear to be processed separately rather than being integrated. Because of its modular neural structure, the cerebellum lends itself readily to computational modeling. It has frequently been suggested that it plays an important role in motor learning (Galea et al., 2011), particularly in error correction (Imamizu et al., 2000; Tseng et al., 2007). The cerebellum makes functional connections with the primary motor cortex, the premotor cortex, and the supplementary motor area. The connections are bilateral; that is, signals are transmitted in both directions. This design allows the cerebellum to monitor command signals to lower motoneurons, providing a means to compare intended actions with sensory feedback about what is actually taking place. It has been suggested that the cerebellum has access to neural representations or internal models of body and environmental dynamics (Imamizu et al., 2003) that allow it to predict the outcome of intended command signals (Miall & Reckess, 2002). Such *forward models* would provide a means of adjusting command signals to lower motoneurons to reduce errors in anticipation of the outcome as opposed to a reaction to the outcome. Forward models will be discussed in more detail in chapter 10.

The basal ganglia receive input from the cerebellum and from many regions of the cerebral cortex. In the context of motor control, inputs from the primary motor cortex, the supplementary motor area, and the primary sensory cortex appear to be predominant. The basal ganglia receive input via two pathways. There is a direct pathway from the cerebral cortex to the striatum and an indirect pathway from the cortex via the subthalamic nucleus. The difference between these two pathways is that the direct pathway removes inhibition of output targets, whereas the indirect pathway reinforces this inhibition. The direct pathway appears to be sharply focused in terms of the actions it controls, whereas the indirect pathway appears to be more diffuse. A current theory is that the indirect pathway acts to prevent undesired movements, allowing the direct pathway to engage the desired movement. A related function that appears to be performed by the basal ganglia is the decision of whether to initiate a particular motor action. However, these ideas have been recently challenged (Turner & Desmurget, 2010) and evidence has been presented that the primary functions of the basal ganglia may be to regulate the magnitude of motor commands and to direct the early stages of motor learning, particularly reinforcement learning.

The motor cortex can be subdivided anatomically and functionally into the primary motor cortex, the dorsal premotor area, the ventral premotor area, and the supplementary motor area. These regions are all involved in planning and execution of actions. The upper motoneurons in these regions make synaptic connections both with interneurons and motoneurons in the spinal cord, although the number of connections is highest for the primary motor cortex. The primary motor cortex is the principal executive region for motor control. Local connectivity within the primary cortex is organized in terms of movements; that is, when a local region of the primary motor cortex is activated, it leads to specific limb movements. A large portion of the primary motor cortex is devoted to the control of finger movements, which is the origin of the extraordinary dexterity of human hands. Although the primary motor cortex is a control center that regulates voluntary motor output to skeletal muscles, it also appears to play an important role in consolidation of motor memory (Galea et al., 2011; Orban de Xivry, Criscimagna-Hemminger & Shadmehr, 2011)—that is, the transition from labile to permanent memory described in chapter 7.

There are extensive functional connections between the supplementary and premotor areas and the primary motor cortex that are likely important in movement planning and preparation. The premotor areas appear to perform sensory transformations between visual and body-referenced coordinate systems. The ventral premotor cortex is involved in mapping sensory information to motor output, such as selecting movements based on the location or shape of a target object. It may also play a role in the selection of movements based on the context of the action. For example, different muscle activation patterns might be required if the action is performed by an individual alone or in cooperation with another person or if it is performed with the aid of a tool. The dorsal premotor cortex represents the movement goal, as opposed to steps to get to the goal, which are represented in primary motor cortex. It integrates information necessary to specify movement parameters such as direction, extent, and speed of movement. The supplementary motor area may be involved in the motivation to act or to select a particular action (Nachev, Kennard & Husain, 2008; Scangos & Stuphorn, 2010)

There are also extensive functional connections between the primary somatosensory area and the primary and premotor regions of the motor cortex. The primary somatosensory area receives input from peripheral sensory receptors, including receptors located in the skin responsible for tactile sensation and those in muscle responsible for proprioception and kinesthesia. This input is integrated and processed to represent body configuration, hand shape, properties of manipulated objects (weight, size, shape, texture), and so on. The processed information is provided as feedback to regions whose output controls the activation of lower motoneurons. The feedback is used to modify ongoing motoneuron activity, providing a mechanism for adapting to changing conditions and for correcting movement errors. The secondary somatosensory cortex receives input from sensory receptors indirectly via the primary somatosensory cortex and is functionally connected to all regions of the motor cortex except the dorsal premotor cortex, as well as the posterior parietal area. Because its input has been preprocessed by the primary somatosensory cortex, its output likely provides a higher-level—that is, more global—representation of the state of the body and the properties of manipulated objects than the output of the primary somatosensory area.

The posterior parietal area is located adjacent to the somatosensory cortex. It is an integrating region that receives a variety of sensory inputs and has extensive outputs to the premotor areas. It receives both proprioceptive and visual information. Certain regions appear to represent the location of an object that is the target of reaching in terms of eye coordinates only, that is, position on the retina. Other regions appear to also represent the location of an object relative to the hand. The posterior parietal cortex is involved in planning and decisions on movement (parallel processes that occurs in several brain areas simultaneously). Decisions are influenced by expectations that are based on past experience and the relative reliability of different forms of sensory information. There is evidence that actions follow a model similar to Bayesian integration that may be encoded in the posterior parietal cortex. Bayesian models of sensorimotor processes are discussed in chapter 10. There is also evidence that a forward model of the impending movement may be represented in posterior parietal cortex, which would facilitate prediction of the current state of the arm for online correction of motor error (Andersen & Cui, 2009).

2.5 Summary

In this chapter, we have presented an overview of fundamental physiological processes involved in the transmission of sensory information in the human nervous system and the generation of commands to muscles for controlling movement, including the molecular basis of action potential generation and conduction and the transmission of information from one neuron to another at chemical synapses by means of synaptic integration. We introduced the concept of a motor unit, which is the fundamental unit of control in the neuromuscular system. We described the structure and function of peripheral sensory receptors responsible for providing tactile, proprioceptive, and kinesthetic feedback to the nervous system, including rapidly and slowly adapting cutaneous receptors, muscle spindles, and Golgi tendon organs. Finally, we briefly described the sensorimotor function of the spinal cord, brainstem, and brain, including the functional specialization of different regions in the sensorimotor network of the brain.

3   Muscle Mechanics and Control

Muscles are biological motors that actively generate force and produce movement through the process of contraction. We begin this chapter by examining the structure and function of muscle at the microscopic level and building to the macroscopic level of whole muscles. This discussion will provide an understanding of the basis for the mechanical behavior of muscle that we observe on the macroscopic level. A great leap in understanding the mechanism of muscle contraction was made when H. E. Huxley and A. F. Huxley independently proposed the sliding-filament cross-bridge model of muscle contraction in 1957 (II. E. Huxley, 1957; A. F. Huxley 1957). The original model has been modified over the years, but the basic principles appear to be sound. We use this model as the basis for understanding how viscoelastic properties of muscle arise. The chapter begins with an explanation of the mechanics of muscle contraction, which is followed by description of the process by which electrical control signals are transformed into muscle force. The final section examines the dependence of muscle force and viscoelasticity on muscle architecture—that is, how musculoskeletal geometry and connective tissue properties affect overall muscle mechanics as manifested at the points of attachment to the skeleton.

3.1   The Molecular Basis of Force Generation in Muscle

Muscle fibers contain two principal contractile proteins, *myosin* and *actin*, whose combined interaction with *adenosine triphosphate* (ATP) converts chemical energy into mechanical work. A myosin molecule consists of a rod (or tail) region and two elongated globular head regions. Myosin rods have self-association properties; that is, they bind to each other to form longer filaments. Myosin filaments are organized such that adjacent pairs of heads are 14.3 nm apart. The filament is mirror symmetric; that is, the head regions along each half of the filament project in opposite directions. Actin molecules are globular proteins with high-affinity binding sites for myosin heads. The globular actin molecules polymerize in the form of long thin filaments.

The actin and myosin filaments in a muscle fiber are arranged in an orderly hexagonal array with six actin (thin) filaments surrounding each myosin (thick) filament (figure 3.1).

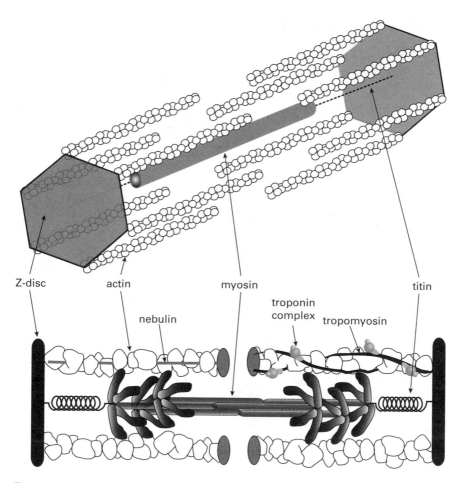

Figure 3.1
A sarcomere contains hexagonal arrays of thin actin filaments surrounding thick myosin filaments. Muscle contraction is regulated by troponin and tropomyosin. Titin contributes to muscle force when stretched and acts to restore relaxed sarcomeres to their resting length (two-dimensional longitudinal section adapted from http://www.bms.ed.ac.uk/research/others/smaciver/Myosin%20II.htm with permission from Sutherland Maciver).

Three pairs of myosin heads occupy each position rotated by approximately 120° with respect to one another around the circumference of the thick filament. Each pair of heads projects toward a different actin filament. Successive positions along the thick filament are occupied by pairs of myosin heads that are rotated by 60° with respect the previous group of three, allowing the thick filament to interact with all six actin filaments that surround it.

Tens to hundreds of thousands of parallel filaments are contained within the cross-section of a single muscle fiber. Along the length of each muscle fiber is a repeating pattern generated by an alternation of partially overlapping arrays of actin and myosin filaments. This repeating unit, known as the *sarcomere* (figure 3.1), is the fundamental force generating unit of muscle. Its relaxed length is about 2–3 μm.

At either end of the sarcomere is a connecting structure that mechanically links the actin filaments of one sarcomere with the next, called the *Z-disc*. Actin filaments project from the Z-discs at either end of the sarcomere toward the center. A large protein molecule called *titin*, anchored at the Z-disc, connects to the myosin filament. It has a region that can be stretched and likely contributes to the passive elastic force that allows a muscle fiber to return to its original length after being stretched.

The actin molecules are not in precise register with the myosin heads. However, if an actin molecule is sufficiently close to a myosin head, a bond can be formed linking the actin and myosin filaments. Because of their ability to form these links, myosin heads are often referred to as *cross-bridges*. The process of formation of a link between actin and myosin is known as *cross-bridge attachment* (figure 3.2). *Cross-bridge detachment* is the reverse process in which the link is broken.

Myosin molecules have ATPase activity; that is, they can bind ATP and then split (hydrolyze) the bound ATP into *adenosine diphosphate* (ADP) and inorganic phosphate (P_i). In the absence of ATP, actin and myosin molecules bind spontaneously to form an actomyosin complex that has a lower free energy than either molecule separately; that is, energy is liberated during the binding of a myosin head to actin. Unless ATP is added to the solution, actin and myosin will remain bound.

ATP can bind to either myosin or actomyosin. The free energy of actomyosin-ATP is similar to that of myosin-ATP, allowing myosin-ATP to detach from actin. The process of active force generation is generally considered to constitute a cycle during which a myosin cross-bridge binds to actin and then later detaches after hydrolyzing ATP. During attachment, the molecular conformation of the myosin cross-bridge changes in such a way that tension is created, pulling the actin filament toward the myosin filament (figure 3.2).

It is generally accepted that the myosin head contains a flexible region that can bend, permitting the portion of the head that is most distant from the actin attachment site to move through a considerable angle. This bending is transformed into linear motion of the actin filament by the lever arm action of the rigid portion of the head (Geeves & Holmes, 1999).

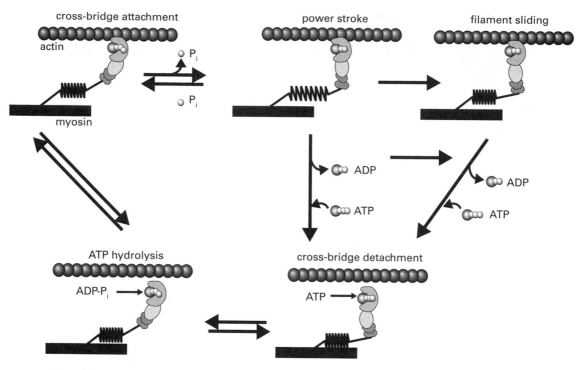

Figure 3.2
During the cross-bridge cycle, myosin heads attach to actin binding sites and undergo a change in conformation powered by ATP hydrolysis to ADP, which causes an increase in tension between the actin and myosin filaments. If the tension in the cross-bridge is greater than the force-resisting movement, the actin filament will slide towards the center of the sarcomere. Following the release of ADP, binding of ATP to myosin initiates cross-bridge detachment (adapted with permission from Nyitrai & Geeves [2004]).

If the actin filament is held isometric so that it cannot move, the cross-bridge applies tension to the actin filament. If the actin filament is free to move, the cross-bridge will pull the actin filament. Both of these processes were recently studied in an experiment that allowed the action of single cross-bridges to be measured. The force produced by a single cross-bridge is several pN; the maximum displacement is on the order of 10–15 nm.

Experiments have been conducted to investigate the relationship between the force and displacement produced by a single cross-bridge moving an elastic load. The force-displacement relation is highly linear, suggesting that the development of tension and filament sliding is powered by the release of elastic energy stored in the cross-bridge.

The mobility of the myosin heads is such that they rotate toward the center of the filament. Consequently, when they attach to actin binding sites, they pull the actin filaments from each half of the sarcomere toward the center, reducing the distance between Z-discs, that is, the sarcomere length. The successive attachment, pulling, and detachment of cross-bridges in an asynchronous manner produces sliding motion of the myofilaments with respect to each other, resulting in muscle shortening, also known as *muscle contraction*.

This cycle can be repeated many times during sustained contraction of a muscle fiber. The cycle period is estimated to be 100–200 ms in duration during isometric contraction, although this value depends on factors such as the myosin ATPase reaction rate, the temperature of the muscle, and the rate of change of muscle fiber length. In particular, at maximum shortening velocity, cross-bridges cannot remain attached for more than 5 ms.

Attachment and force generation, and detachment, may proceed much faster than the overall cycling rate. The attachment and force generation step has been estimated to occur within as little as 7 ms; the detachment step can occur within 2.5 ms. Myosin exists in a number of isoenzymic forms that differ in their rates of splitting ATP. These different isoenzymes are important in determining the maximum shortening velocity of a sarcomere.

The amount of force that a sarcomere can produce depends on a number of factors, of which sarcomere length, rate of change of length, and recent mechanical and activation history are arguably the most critical. The classical textbook picture of the dependence of sarcomere force on length is one in which the force is proportional to the amount of overlap between thick and thin filaments (Gordon, Huxley & Julian, 1966). This relation is logical because sarcomere force should reflect the number of bonds that can be formed between actin and myosin, that is, the number of cross-bridges. Starting from a long sarcomere in which there is no overlap, a linear increase in sarcomere force occurs as the sarcomere length is reduced. A point of maximum force is reached when there is complete overlap between the thick and thin filaments. However, sarcomere force can be considerably enhanced or depressed (figure 3.3) depending on whether the sarcomere is stretched or shortened prior to reaching its final length (Herzog, Joumaa & Leonard 2010). There is some evidence that these effects arise from the mechanical properties of the titin molecule. At sarcomere lengths shorter than the point of maximum overlap, the force is expected to decline for several reasons, including interference between thin filaments attached to opposite ends of the sarcomere and compression of titin molecules and thick filaments at very short lengths.

Muscle force is generally separated into *active* and *passive* components. Passive forces are generated when structural proteins that maintain the shape and integrity of a muscle are stretched, which includes stretching of cross-bridges. Active forces are those generated by a change in molecular conformation that requires energy expenditure, such as cross-bridge bending driven by ATP hydrolysis. The total force generated by a muscle is the sum of passive and active forces.

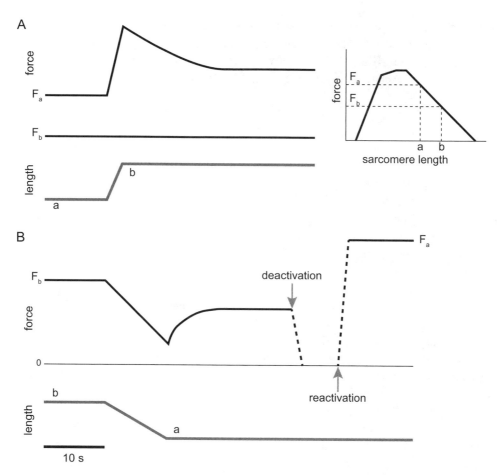

Figure 3.3
(A) When a sarcomere, initially at length **a**, is stretched to length **b** after being activated, its force after stretching is enhanced relative to the force it would have generated at length **b** had it not been stretched. (B) When a sarcomere, initially at length **b**, is shortened to length **a** after being activated, its force at the new length is depressed relative to the force it would have generated at length **a** (F_a) had it not been stretched, that is, after deactivation and reactivation.

3.2   The Molecular Basis of Viscoelasticity in Muscle

Cross-bridges behave like tiny springs (figure 3.2). If the spring is stretched by pulling on one end, the tension in the spring increases. If the spring is shortened, the tension in the spring decreases. An attached cross-bridge can be stretched or shortened by increasing or decreasing the length of the sarcomere in which it is located. *Stiffness* is defined as the ratio of the change in force to the change in length:

$$K = \frac{dF}{dx} \tag{3.1}$$

Sarcomere stiffness is frequently modeled as the slope of the relation between isometric force and sarcomere length. Such models are erroneous because they fail to account for the magnitude of the change in force when a sarcomere is transiently stretched or shortened by a small amount. This force reflects the number of attached cross-bridges. In contrast, the slope of isometric force-length relation remains constant as the number of attached cross-bridge decreases linearly with length. Furthermore, the slope of this relation is negative; that is, isometric force drops as length increases beyond the point of maximum overlap and there is abundant evidence that muscles do not exhibit characteristics of negative stiffness.

In the case of n elastic elements, such as cross-bridges, acting in parallel, we have summation of forces. A length change dx applied to a parallel arrangement of elastic elements causes the length of each element to change by dx and hence its force changes by an amount Kdx, where K is its stiffness. Thus, for elastic elements acting in parallel, we have

$$dF = dF_1 + dF_2 + \cdots + dF_n = K_1 dx + K_2 dx + \cdots + K_n dx$$
$$K = \frac{dF}{dx} = K_1 + K_2 + \cdots + K_n \tag{3.2}$$

Because all attached cross-bridges are stretched when a sarcomere is stretched, they all contribute to the force applied to the thin filaments. All of these forces are applied in parallel to the thin filaments so they sum. If there are N_c attached cross-bridges, which are each stretched by an amount dx_s and each cross-bridge has stiffness K_c, then the change in sarcomere force is given by

$$dF_s = N_c K_c dx_s \tag{3.3}$$

It follows that the sarcomere stiffness is proportional to the number of attached cross-bridges, that is,

$$K_s = N_c K_c \tag{3.4}$$

Although stiffness can obviously be modulated by a change in the number of cross-bridges, it appears from recent studies that other mechanisms may also contribute to stiffness modulation. One of these involves the formation of chemical bonds between actin and titin. These bonds would effectively act in parallel with the stiffness of actin filaments and titin, increasing sarcomere stiffness.

A muscle fiber is composed of a collection of parallel *myofibrils*, each of which consists of several thousand sarcomeres joined end to end in series. A change in the length of a myofibril must be equal to the sum of the length changes in the individual sarcomeres. However, the change in force must be the same for each sarcomere. This relationship can best be illustrated by considering two neighboring sarcomeres. If the change in force of one sarcomere were greater than in the neighboring sarcomere, it would pull on the neighboring sarcomere. This pull would stretch the attached cross-bridges in the neighboring sarcomere until the forces in the two sarcomeres balanced. Thus, the force generated by each sarcomere would be equal after the length change.

In the case of a chain of elastic elements in series, such as sarcomeres, the change in force must be the same everywhere in the chain, that is, $dF = dF_1 = dF_2 = \cdots = dF_n$. However, the length change is distributed among the elements such that the sum is equal to the overall change in length. Thus, for elastic elements acting in series, we have

$$dx = dx_1 + dx_2 + \cdots + dx_n = \frac{dF_1}{K_1} + \frac{dF_2}{K_2} + \cdots + \frac{dF_n}{K_n} = dF\left(\frac{1}{K_1} + \frac{1}{K_2} + \cdots + \frac{1}{K_n}\right)$$

$$K = \frac{dF}{dx} = \frac{1}{\dfrac{1}{K_1} + \dfrac{1}{K_2} + \cdots + \dfrac{1}{K_n}} = \frac{\displaystyle\prod_{i=1}^{n} K_i}{\displaystyle\sum_{i=1}^{n}\left(\prod_{j\neq i} K_j\right)} \tag{3.5}$$

It follows that for a myofibril consisting of N_s sarcomeres of stiffness K_s,

$$K_m = \frac{K_s^n}{N_s K_s^{n-1}} = \frac{K_s}{N_s} \tag{3.6}$$

From this relation, it can be seen that the greater the number of sarcomeres, N_s, the lower the stiffness of a myofibril. It follows that the longer the myofibrils, the less stiff the muscle fiber.

As might be expected, when a muscle fiber is activated to produce a steady force while being held isometric and is then stretched at constant velocity, the resulting force is greater than the isometric force (figure 3.4). Although the force increases with velocity for low velocities of stretch, it levels off as the velocity increases further, never reaching more than about 1.5 times the isometric force. The increase in force with muscle lengthening velocity is probably largely due to stretching of attached cross-bridges. Cross-bridges that are being stretched will generate a greater average force during their period of attachment than cross-bridges that are isometric. The higher the lengthening velocity, the more cross-bridges are stretched during the period of attachment. As the lengthening velocity increases,

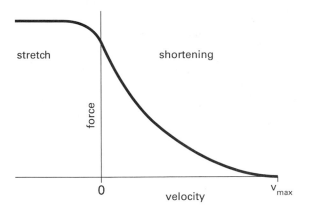

Figure 3.4
When a sarcomere is being stretched (negative velocity), its force is greater than the isometric force (zero velocity). When a sarcomere is being shortened, its force drops below the isometric force, decreasing inversely with the shortening velocity.

cross-bridges will be stretched beyond the limits that can be supported by the binding force between actin and myosin, resulting in forcible detachment. This behavior limits the maximum force during muscle lengthening.

When a sarcomere shortens against a load, the velocity of shortening automatically adjusts to the size of the load. As the load increases the shortening velocity decreases (Edman, 2010). From the point of view of velocity, force decreases in a hyperbolic fashion as the velocity of shortening increases (figure 3.4). Force decreases continuously from its isometric value to zero at maximum shortening velocity, although the relation deviates slightly from a smooth hyperbola near zero velocity. There are several possible reasons why sarcomere force drops as the velocity of shortening increases. First, there are fewer cross-bridges attached during shortening and their number decreases as the velocity of shortening increases. It has been suggested that this is a consequence of an increase in the rate of cross-bridge detachment during muscle shortening and a decrease in the rate of attachment. Both of these rates may be functions of velocity. Second, shortening reduces the tension in attached cross-bridges (figure 3.2). Cross-bridges that are shortening will generate a smaller average force during their period of attachment than cross-bridges that are isometric. At higher shortening velocity, greater shortening may occur during the period of attachment resulting in lower average force during the period of cross-bridge attachment. Third, some cross-bridges may be compressed as the result of shortening before they detach. These cross-bridges would generate negative force, thereby reducing the overall tension developed by the sarcomere. The higher the shortening velocity, the more quickly cross-bridges would compress, resulting in a greater number of cross-bridges generating negative force before detachment.

Whenever a cross-bridge detaches, it loses all of its *elastic energy*. This energy is dissipated. A mass-spring system that dissipates elastic energy has the property of *damping*; that is, when it oscillates, the amplitude of the oscillations decreases as elastic energy is dissipated. In the case of a sarcomere, this damping depends on the velocity. A system that resists a change in velocity by producing a change in resistive force has the property of *viscosity*. The *coefficient of viscosity* is defined as the change in force produced by a change in velocity, $\dot{x}$:

$$D = \frac{dF}{d\dot{x}} \tag{3.7}$$

In the case of a sarcomere, resistive force always pulls toward the center. By convention, sarcomere lengthening velocity is generally represented as negative and sarcomere shortening velocity is positive. However, to be consistent with the conventions of Newtonian mechanics, we will represent muscle lengthening velocity as positive and muscle shortening velocity as negative. Sarcomere force is positive when the force pulls the ends of the sarcomere toward the center. An increase in sarcomere lengthening velocity near zero produces a positive change in sarcomere velocity and a positive change in force; that is, sarcomere force increases. This change results in a positive value for the coefficient of viscosity, D. If a sarcomere is shortening and the shortening velocity increases, the force will decrease. This change also results in a positive value for D because both the change in force and the change in velocity are negative. Sarcomere viscosity is highest around zero velocity and lowest at high lengthening and shortening velocities where there is little change in force. Like stiffness, damping depends on the number of attached cross-bridges; that is, the amount of elastic energy that can be dissipated increases with the number of attached cross-bridges.

3.3 Control of Muscle Force

The formation of cross-bridges is largely regulated by *tropomyosin* and *troponin* molecules associated with actin filaments (Gordon, Homsher & Regnier, 2000). Tropomyosin forms strands that overlay actin filaments in a way that interferes with the myosin binding. A troponin complex is bound to each tropomyosin molecule. A troponin complex consists of troponin T, I, and C. Troponin T binds the complex to tropomyosin; troponin I binds to actin and inhibits the ATPase activity of actomyosin. Troponin C has two high-affinity Ca^{2+} binding sites that participate in binding of troponin to actin and two low-affinity Ca^{2+} binding sites that are involved in regulation of contraction.

In the absence of Ca^{2+} binding to troponin C, the tropomyosin blocks interaction of myosin with actin binding sites. Ca^{2+} binding to low-affinity binding sites on troponin C causes a *conformational change* that moves the tropomyosin molecule away from the myosin-binding site on actin (figure 3.5). This movement allows binding of myosin

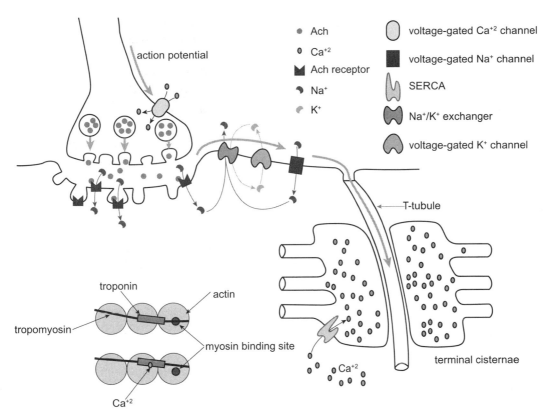

Figure 3.5
When an action potential reaches an axon terminal, it triggers release of acetylcholine, which diffuses across the
synaptic cleft and binds to receptors on cation channels. Opening of the cation channels leads to initiation of an
action potential that propagates along the membrane of the muscle fiber into the T-tubules, triggering release of
Ca^{2+} from the terminal cisternae into the myoplasm. Muscle contraction is regulated by binding of Ca^{2+} to troponin, which leads to a shift in the position of the tropomyosin filament, exposing myosin binding sites on actin
filaments.

cross-bridges in the strong (force-generating) state. Weak cross-bridge binding can occur
in the absence of Ca^{2+}. Although cross-bridges in the weak binding state do not actively
exert force, they do contribute to the stiffness of a sarcomere.

As described in chapter 2, action potentials are conducted from the motoneuron soma,
located in the spinal cord, to the muscle fibers along axons. At the axon terminal, a trans-
mitter molecule is released and diffuses across the synaptic cleft, where it binds to a
receptor in the postsynaptic membrane of the muscle fiber (figure 3.5). In the process of
binding to the receptor, an ion channel opens that allowing Na^+ to move from the extracellular to the intracellular fluid and K^+ to move from the intracellular to the extracellular
fluid. This movement results in a net depolarization of the postsynaptic membrane that is

sufficient to trigger opening of voltage-gated Na^+ channels and initiate an action potential. The action potential propagates from the neuromuscular junction along the plasma membrane of the muscle fiber and reaches the interior of a muscle fiber through *transverse tubules* (T-tubules), which form a network that penetrates the muscle fiber. The depolarization of the T-tubule membrane leads to opening of voltage-gated Ca^{2+} channels, which allows Ca^{2+} to diffuse from storage sites in the *sarcoplasmic reticulum* into the *myoplasm* of a muscle fiber (Dulhunty, 2006).

After its release, Ca^{2+} quickly diffuses and rapidly binds to the troponin C, allowing strong cross-bridge binding to take place. As a consequence of cross-bridge formation, active force will be generated and muscle shortening will occur if actin and myosin filaments are free to slide past one another. The number of cross-bridges that can exist is determined by the concentration of free Ca^{2+} in the myoplasm. The greater the concentration of Ca^{2+}, the greater the number of cross-bridges and the greater the sarcomere force. However, there is a limit to the number of cross-bridges that participate in the generation of force at any given time, because even at the highest Ca^{2+} concentrations, only about 30 percent of the total number of myosin heads appears to be attached to actin binding sites.

The concentration of free Ca^{2+} in the myoplasm is rapidly reduced by the action of *calcium pumps* in the sarcoplasmic reticulum following an action potential, as well as by the buffering action of *Ca^{2+} binding proteins* in the myoplasm and the *mitochondria*. As Ca^{2+} is released from troponin molecules, the contractile force quickly declines as the Ca^{2+} concentration in the myoplasm is brought back to its pre-excitation level, allowing the muscle to relax. A single muscle action potential produces a brief contraction of the muscle fiber called a *twitch*. The duration of the twitch depends on the *muscle fiber type*. The duration of both the contraction and relaxation phases of the twitch is longer for *slow-twitch* (type I) than *fast-twitch* (type II) fibers (figure 3.6).

The CNS controls tension by specifying the number of active motor units (*recruitment*) and their firing rates (*rate coding*). The force produced by each muscle fiber of the motor unit increases with its firing rate because of the accumulation of intracellular Ca^{2+}. Each action potential depolarizes the muscle membrane, which results in more Ca^{2+} being released from the terminal cisternae, diffusing through the intracellular space and activating more actin-binding sites. When a motor unit first becomes active, its firing rate is about 8 Hz. Although humans can activate motor units at instantaneous firing rates of up to 100 Hz during brief forceful efforts, the maximum firing rates that they sustain during steady contractions are considerably lower and generally do not exceed 30 Hz. However, these rates are sufficiently high that several action potentials can occur before the twitch force from the first action potential has dropped to zero. Whereas the muscle action potential has a duration of less than 10 ms, the twitch duration for skeletal muscle fibers is of the order of 100–200 ms. Successive action potentials that occur at intervals of less than the twitch duration allow Ca^{2+} to accumulate in the muscle fiber. As the amount of accumulated

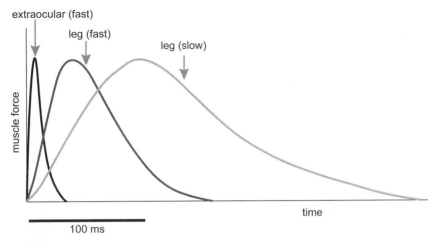

Figure 3.6
The force generated by a motor unit in response to a single action potential is referred to as a twitch. The rate at which twitch force increases and decreases varies with muscle fiber type. Extraocular muscles have the fastest contraction and relaxation times. The slowest contracting skeletal muscle fibers are found in larger postural muscles, which are often continuously active.

Ca^{2+} increases, the force also increases because the number of cross-bridges that can form depends on the amount of Ca^{2+} available for binding to troponin.

When a muscle is activated voluntarily under isometric conditions, motor units tend to become active in a fixed order. The recruitment order is roughly correlated with the amount of force that a motor unit can produce, which in turn depends on the number of muscle fibers and the size of the muscle fibers that it comprises. The motor unit that produces the smallest force tends to become active first (Tansey & Botterman, 1996). As the strength of the motor command increases, additional motor units that produce larger forces become active. A motor unit remains active until the strength of the motor command drops below its *force threshold*. As the strength of the motor command increases, the firing rate of active motor units increases, which results in greater force.

In controlling force, the CNS must allow for the dependence of force on muscle fiber length and velocity, as already described. However, the control problem is even more complex because the force also depends on the past history of activation, length, and velocity. For example, the motor unit force generated at a particular firing rate is less when motor unit activation is maintained at a constant level (firing rate) than when a short interval of higher activation precedes the constant level (figure 3.7). Furthermore, if the muscle fiber is lengthened immediately prior to being activated or while being activated, this force can be higher than when the sarcomere length is held constant prior to and throughout activation (figure 3.3). By the same token, force at a given sarcomere length can be less when the muscle fiber is shortened to a given length than when it is held

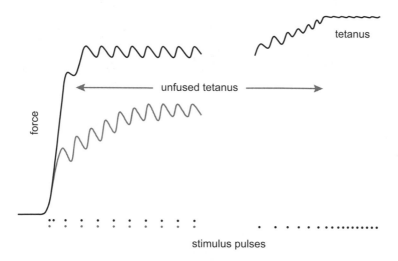

Figure 3.7
The force generated by a motor unit during repetitive activation depends on the activation rate. At low rates of activation, the force is modulated as the muscle fiber contracts and relaxes. As the rate of activation increases, the force increases, but there is less time for relaxation, so the force becomes smooth and achieves its maximum at tetanus.

isometric at that length (figure 3.3). The size of these effects depends on how quickly the muscle length was changed. These examples suggest that without previous rehearsal, it may be practically impossible for the CNS to accurately predict the force that a muscle will generate for a given high-level command. Nevertheless, with frequent repetition, the CNS may be able to consistently recreate a "desired" history, which would ensure similar muscle force each time an activity is performed. However, it also explains why performance can be highly variable despite the intent to be consistent.

3.4 Muscle Bandwidth

The transformation of action potentials into muscle force represents a low-pass filtering process; that is, as the motor unit firing frequency increases, the amplitude of the force modulation decreases because the muscle fiber does not completely relax between successive action potentials (Partridge, 1965). Instead the mean force increases. At sufficiently high firing rates, the force reaches a maximum value and can no longer be modulated. By modulating the firing rate at different frequencies, it is possible to characterize the control bandwidth. For limb muscles the effective cutoff frequency is 2–3 Hz. Thus the ability of a single muscle to modulate force at higher frequencies is severely limited. The delay between release of Ca^{2+} and onset of cross-bridge formation together with the low-pass filtering creates a lag between the electrical input and mechanical output that increases as the frequency of the electrical input decreases. It can range from about 30 ms for rapid contraction to over 300 ms for very slow modulation of muscle activity as occurs during postural sway.

During prolonged muscle activation, there is a reduction in the force producing capacity of a muscle fiber, known as *fatigue*. The peak twitch amplitude declines and twitch duration increases with fatigue. Consequently, muscle force decreases and its response slows; that is, the control bandwidth is reduced with fatigue.

Of course, muscles do not work as isolated units. In particular, they operate as agonist-antagonistic pairs to move a joint in opposite directions. This mechanical organization permits higher bandwidths to be achieved because muscle force can be increased more quickly than it can be reduced. Strong activation of an antagonist muscle group produces a rapid increase in opposing torque at the joint that causes the net joint torque to drop more quickly than it would by relaxation of the agonist muscle group alone.

3.5 Muscle Fiber Viscoelasticity

When a muscle fiber undergoes a length change, the length change is distributed among all of the sarcomeres in the myofibrils that make up the muscle fiber. All of the sarcomeres in a myofibril will undergo the same change in force when their length changes, and this change in force will be equal to the change in force of the myofibril itself, but as implied by equation (3.5), the change in force will depend on the total number of sarcomeres. The more sarcomeres in a myofibril, the smaller the length change of each sarcomere and hence the smaller the change in force arising from the cross-bridge stiffness. The same reasoning can be applied to the force arising from sarcomere viscosity because the rate of change of length (velocity) will also be distributed among the sarcomeres. Thus, muscles that have the shortest muscle fibers—that is, those that are composed of myofibrils with the fewest sarcomeres in series—will produce the greatest force to oppose changes in length or velocity.

When a muscle fiber is stretched, force increases and stiffness remains relatively constant until the sarcomere length has increased by 15–20 nm (Campbell, 2010). At this point, the elastic force created by stretch of a cross-bridge reaches the limit that can be held by the actin-myosin bond. Further stretching causes the actin-myosin bond to break. This behavior defines the limit of a region referred to as *short-range stiffness*. During stretch of a slow-twitch (type I) muscle fiber, the force will often stop increasing or even begin to decrease if the fiber is stretched beyond this point. When this happens, the muscle fiber is said to *yield* (figure 3.8). A clearly marked region of short-range stiffness is not seen in all muscle fiber types. For example, fast-twitch (type II) fibers in the gastrocnemius muscle of the cat may show only a moderate decline in the rate of change of force during stretch (Malamud, Godt & Nichols 1996). This finding suggests that the yielding behavior may be related to ATPase rates, governing the rates of cross-bridge attachment and

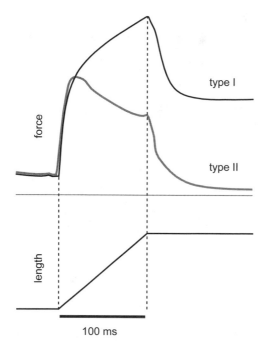

100 ms

Figure 3.8
Slow-twitch motor units (type II) increase force during stretch until reaching the point at which tension in cross-bridges is greater than the actin-myosin binding force. The force stops increasing or yields at this point as cross-bridges are forced to detach. Fast-twitch motor units (type I) do not yield but increase force more gradually beyond the yield point because the cross-bridge attachment rate is much higher than for slow-twitch motor units.

detachment. In type II muscle fibers, cross-bridges will reattach and begin generating force more quickly after forcible detachment than in type I muscle fibers. Consequently, forcible detachment will have less of an effect on the rate of force development in type II muscle fibers.

When a muscle fiber is stretched at low velocities beyond the region of short-range stiffness, the stiffness is found to be slightly higher than in the isometric state. During lengthening, actin filaments move in the same direction as the myosin heads of detaching cross-bridges. This movement reduces the relative velocity of myosin heads with respect to actin binding sites in comparison with the isometric state. At low velocities of lengthening, this behavior may facilitate binding of myosin to actin, creating a greater number of attached cross-bridges than in the isometric state. Muscle fiber stiffness decreases during shortening as shortening velocity increases. The decrease in stiffness during shortening suggests that there are fewer attached cross-bridges than in the isometric state. Once shortening stops, the stiffness and force quickly recover, suggesting reattachment of cross-bridges that had detached while the muscle fiber was shortening.

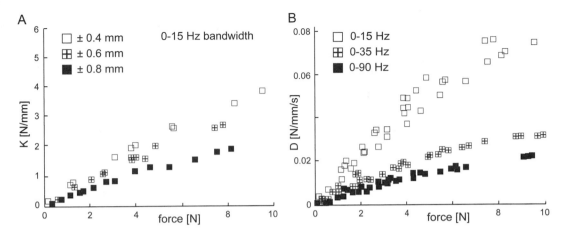

Figure 3.9
(A) Stiffness increases with muscle force but is lower when measured using large changes in muscle length compared to small changes in muscle length. (B) Viscosity also increases with muscle force but is lower when measured using faster rates of muscle length change, that is, higher-frequency length changes (adapted with permission from Kirsch, Boskov & Rymer [1994]).

In addition to muscle fiber length, length change, and rate of change of length, visco-elasticity is a function of the activation level (Kirsch, Boscov & Rymer, 1994). This relationship is evident in figure 3.9, in which stiffness and viscosity are plotted as a function of muscle force. The data suggest that both stiffness and viscosity increase linearly with muscle force. This behavior would be predicted from a linear relation between sarcomere force and number of attached cross-bridges. The data in this figure were obtained for pulsatile displacements that varied pseudorandomly in duration. The results indicate that stiffness decreases as the amplitude of the pulses increases, as would be predicted by short-range stiffness. Furthermore, viscosity decreases for higher frequency bands, which can be attributed to higher average rates of muscle length change.

3.6 Muscle Geometry

Muscle fibers are linked to each other and to the skeleton by collagenous connective tissue. They do not attach directly to bone but apply forces to the skeleton through sheets of connective tissue called aponeurosis and collagen bundles that form tendons, which are attached to bones. The proximal site of attachment of the tendon is known as its *origin* and the distal attachment site is its *insertion*. The origin and insertion sites are located on opposite sides of a joint. Depending on the type of joint, muscle contraction produces either rotation or sliding of the joint. In some cases, such as the fingers, a tendon can cross several joints.

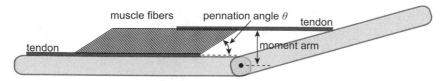

Figure 3.10
In most muscles, muscle fibers are pennated; that is, they attach at an angle to tendons or aponeurosis. The angle of pennation, *θ*, can vary significantly with joint angle and with muscle force because muscle fibers shorten, stretching the tendon during muscle contraction. The perpendicular distance between the line of action of the tendon and the center of rotation of the joint is known as the moment arm.

The geometrical arrangement of the fibers and connective tissue plays an important role in determining muscle force and mechanical behavior (Lieber & Ward, 2011; Gans & Gaunt, 1991). The shape of the tendon affects its physical properties and also determines how the muscle fibers can be geometrically arranged. Muscle fibers generally attach to aponeurosis or tendon in parallel arrays (figure 3.10). In almost all cases, the muscle fibers attach at a slight angle with respect to the line of pull of the tendon. This angle is called the *pennation angle*.

The pennation angle of a muscle at rest length varies considerably among muscles. Some muscles have been shown to have pennation angles as small as 10° under resting conditions at neutral joint angles, whereas others may have pennation angles as large as 20–25°. Pennation angle increases as the length of the *muscle-tendon unit* is shortened by rotation of the joint. If a muscle contracts or if its length is changed while it is contracting, there can be even greater variations in pennation angle. In some muscles, the pennation angles may be as large as 50° when the muscles are contracting maximally at short lengths, whereas others show almost no change in pennation angle.

The relevant parameter for determining the force-generating capacity of a muscle is its *physiological cross-sectional area* (PCSA). The PCSA is the cross-sectional area measured perpendicular to the long axes of the muscle fibers. It is not the same as the anatomical cross-sectional area, which is perpendicular to the muscle. The PCSA is a measure of the number of sarcomeres working in parallel within a muscle.

The force vector generated by a pennate array of muscle fibers has a component that lies along the line of action of the tendon that contributes to force and motion at the origin or insertion sites. The other component is perpendicular to the line of action of the tendon and makes no contribution to force along the line of action of the tendon. The greater the pennation angle, the less force a muscle fiber contributes along the line of action. The component of the muscle fiber force perpendicular to the line of action of the tendon causes muscle fibers to push against each other, against other soft tissue, or against bone. This movement leads to changes in the shape of the muscle during contraction. These changes in shape may cause the tendon to shift position, leading to a change in the direction of the line of action with respect to the site of origin or insertion during contraction.

In *multipennate* muscles, the fibers may have a complex geometrical arrangement with sheets of fibers oriented in different planes. During contraction, local rotations and deformations of sets of fibers will occur in each plane, leading to complex changes in shape. With such complex arrangements of aponeuroses, muscle fiber lengths may vary from one region of the muscle to another in addition to being oriented in different planes. This variation can lead to different sarcomere lengths for muscle fibers in different regions of the muscle. Because the length-tension relation for the whole muscle is an average of the length-tension relations of all of its fibers, the shape of the length-tension curve for the whole muscle is not simply a scaled version of the sarcomere length-tension curve.

Some muscles, such as the trapezius, the deltoid, and the latissimus dorsi have triangular arrangements of muscle fibers. In these cases, muscle fibers radiate in different directions from the origin to insertion sites. Muscle fibers originate from a small area from which they fan out to a broad insertion region. The lines of action of such muscles can be modified by activating different populations of muscle fibers, permitting the same muscle to apply force and produce movements through a range of directions.

In *fusiform* muscles such as the biceps, fibers run parallel to one another along the entire length of the muscle. Midway between origin and insertion, the fibers form a cylinder. However, because the cross-sectional area of the tendon is much less than the cross-sectional area of the central region of the muscle, the fibers must taper considerably as they approach the tendon junction. Consequently, the outer fibers will follow a curved path from origin to insertion. When the fibers in the center of the muscle are activated, they tend to shorten and bulge, which displaces adjacent fibers laterally, causing the peripheral fibers to curve farther from the central axis of the muscle. Thus, the most peripheral fibers can have large differences in their pulling directions from one end of the fiber to the other.

3.7   Tendon Mechanics

When tendon is stretched from a relaxed (unloaded) state, there is initially very little resistance to stretch, as is indicated by a concave region in the curve of force plotted against elongation, termed the *toe region*. Continued elongation within the toe region produces an increase in the stiffness of the tendon, necessitating greater forces for equivalent elongation. Elongation beyond the toe region produces a linear increase in force up to the point of tendon failure (figure 3.11).

Tendons vary considerably in shape, ranging from flat bands to cylindrical cords. They may be short and thick or long and thin. Tendon stiffness increases as the tendon is stretched. Tendon stiffness is very low below about 1% strain, then begins to increase until the strain reaches about 4 percent, beyond which it remains relatively constant (Maganaris & Paul, 2002). Tendon failure occurs at about 8–10 percent strain; that is, tendons will rupture if they are stretched by more than 10 percent of their resting length. In some muscles, it would appear that the stiffness continues to increase over the entire range of

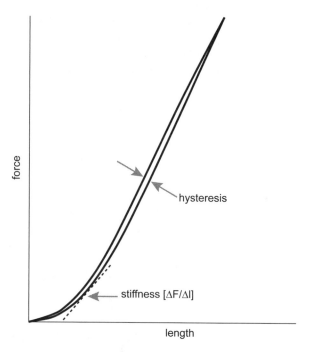

Figure 3.11
When tendon is stretched from a relaxed (unloaded) state, there is initially very little resistance to stretch; that is, its stiffness is low. Its stiffness increases and becomes relatively constant at greater lengths, that is, at higher strain. When the tendon is shortened, the force at a given length is slightly lower than when it is lengthened. This hysteresis represents lost elastic energy.

force (Sugisaki et al., 2011). The tendon stiffness K_t is proportional to the cross-sectional area, A, and inversely proportional to the tendon length, x_t:

$$K_t \propto \frac{A}{x_t} \tag{3.8}$$

Tendon stiffness increases with thickness and decreases with length. Thicker tendons are stiffer and longer tendons are more compliant.

When a tendon is stretched and then brought back to its original length, the force-displacement curve follows a different path during relaxation than it does during stretch. The shape of the curve is similar but is displaced slightly toward lower forces; that is, for the same tendon length (or strain), the force is lower (figure 3.11). This difference is due to dissipative forces in the tendon. The separation between these two curves is known as *hysteresis*. The area contained between the boundaries of the elongation and relaxation curves represents the energy put into the tendon, which is not returned on recoil; the greater the hysteresis, the greater the loss of stored energy. The energy loss in a tendon due to hysteresis is typically less than 10%, indicating that its elastic properties dominate over its dissipative properties. Muscle fibers have much greater hysteresis than tendons because of the large velocity-dependent difference in force that is independent of muscle length (figure 3.4).

3.8 Muscle-Tendon Unit

Because tendon and aponeurosis are elastic elements in series with muscle, the relative stiffness of the tendon or aponeurosis compared to muscle can profoundly affect the mechanical behavior of the muscle-tendon unit. Under resting conditions when no force is being applied to the tendon, tendon, aponeurosis, and muscle fibers may all be relatively compliant. When the muscle fibers begin to contract, they pull on the tendon and cause it to stretch (Ito et al., 1998). This stretching acts like a low-pass mechanical filter between the muscle fibers and the skeleton, causing a lag in the application of muscle force to the skeleton. The more compliant the tendon, the more it will stretch and the greater the lag before force is applied to the skeleton.

As the muscle force increases during contraction, the tendon will stretch and generally become stiffer. Muscle fibers will also become stiffer as the level of muscle activation increases because the number of attached cross-bridges increases with activation. If activation is sufficiently high, muscle stiffness can exceed tendon stiffness. It is often assumed that any movement of the joint that increases the length of the muscle-tendon unit results in stretching of both the muscle fibers and the tendon. However, muscle imaging studies have shown that in some cases, only the tendon stretches, whereas muscle fibers appear to shorten likely because the externally applied force cannot overcome a combination of high resistance to stretch and large contractile force. In some ways, this is an energy efficient design, because tendons store elastic energy much better than muscles, which tend to dissipate a significant fraction of the elastic energy stored in cross-bridges when stretched.

Muscle fibers and tendon constitute a series arrangement of elastic elements. Applying equation (3.5), the overall stiffness of the muscle-tendon unit is given by

$$K = \frac{K_\mu K_t}{K_\mu + K_t} \tag{3.9}$$

where K_μ is the stiffness of the muscle fibers and K_t is the tendon stiffness. Clearly, the stiffness of the muscle tendon unit is always less than that of either the muscle fibers or the tendon.

3.9   Summary

In this chapter, we have presented the molecular basis of the generation of force and motion in muscle, the biological motor of the human body. We have explained some of the non-linear properties of muscle force generation and the implication for control of movement. We have emphasized that muscle acts not only as a force generator but also as an element with viscoelastic properties, which—like force—are under the control of the CNS. We have described different muscle geometries and how they differ functionally, highlighting the features of pennate muscles. Finally, we considered the mechanical properties of tendon, which—unlike muscle fibers—are excellent elastic storage elements. We made the point that ultimately the mechanics of the muscle-tendon unit—that is, the serial arrangement of muscle fibers and tendon, as opposed to the mechanics of isolated muscle fibers—must be considered the fundamental model of musculoskeletal mechanics.

4 Single-Joint Neuromechanics

In this chapter, the dynamics of a single joint under the control of the CNS are described. A *single joint* in the context of this chapter is a single *axis of rotation* about which a body segment rotates, such as forearm rotation about the elbow. There are a number of aspects to consider, including the mechanics of the body segments on either side of the joint, the mechanics of the muscles attached to the skeleton on either side of the joint, and the neural circuits that control the muscles. Although a single joint will be considered in isolation, it is important to keep in mind that each joint works in concert with the joints to which it is linked and that the motion of those joints must be simultaneously controlled. In particular, the CNS must account for reaction forces and torques on these joints to ensure that their movements are properly coordinated or that they are appropriately stabilized to achieve a desired movement. The coordinated control of multiple joints is discussed in chapters 5 and 6.

4.1 Joint Kinematics

Joint function is determined primarily by the shape and contour of the contact surfaces and constraints of the surrounding soft tissue. A *hinge joint*, or revolute joint, is the simplest but most common model used to simulate an anatomical joint in planar motion. Movement of a body segment is confined to one plane about a single axis of rotation. In general, hinge motion is used to describe movement of body segments that are constrained to move in a single plane, such as forearm flexion and extension about the elbow or flexion and extension of the phalanges of the fingers. Articulation of the ulna about the radius, which involves rotation of an arch-shaped surface about a rounded or peg-like pivot, can also be modeled as hinge motion (figure 4.1A).

A more general planar joint allows translation of the center of rotation in a plane. This joint has three degrees of freedom (two translations and one rotation). Usually, this motion consists of gliding movement such as that between the carpal bones of the wrist. It is also often used for the knee, where the center of rotation translates during movement. A *universal joint* allows rotation about two axes through the joint and has two degrees of

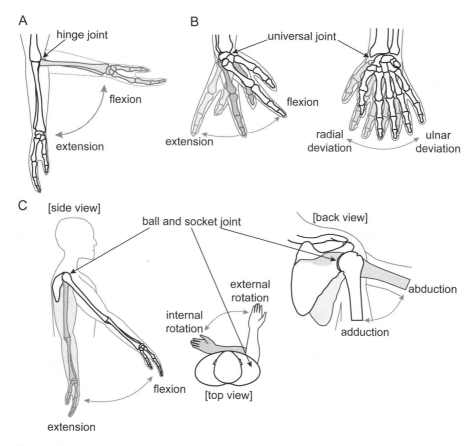

Figure 4.1
(A) The elbow is an example of a joint with one degree of freedom for rotational movement that acts like a hinge joint. (B) The wrist has two degrees of freedom and acts like a universal joint. (C) The glenohumeral joint of the shoulder has three degrees of freedom and acts like a ball and socket joint.

freedom. The universal joint is associated with joints in which a saddle-shaped bone fits onto a socket that is concave-convex in the perpendicular directions, such as the articulation of the first metacarpal and trapezium. It also describes the movement of joints such as the wrist joint, where movement takes place in two perpendicular planes (figure 4.1B). A *ball and socket joint* consists of a ball-shaped head that fits into a concave socket where the body segment rotates about three axes that intersect at the joint center. This joint has three degrees of freedom and is commonly used to model three-dimensional joint movement such as occurs at the glenohumeral joint of the shoulder or the hip joint (figure 4.1C).

In general, there is some translation of the articulated surfaces of joints, in addition to rotation. However, it is still possible to describe the motion as rotation in these cases by defining an instantaneous axis of rotation that shifts slightly during the movement. These shifts in the location of the axis of rotation may be insignificant for qualitative description of movement but more important for the estimation of muscle force.

4.2   Joint Mechanics

The *moment arm* of a muscle is the perpendicular distance from the line of action of the muscle to the instantaneous axis of rotation of the joint (figure 3.10). The moment arm determines the transformation of muscle force to joint torque and muscle motion to joint motion: the longer the muscle moment arm, the greater the joint torque for a given muscle force, but the less the joint moves for a given amount of muscle movement. This is evident from equation (4.1):

$$\tau = \rho\mu$$
$$\rho = \frac{d\lambda}{dq} \Rightarrow dq = \frac{d\lambda}{\rho} \tag{4.1}$$

where τ is the joint torque, μ is the muscle force, q is the joint angle, λ is the muscle length, and ρ is the muscle moment arm.

The length of the muscle moment arm often changes as joint angle changes, primarily due to change in direction of the line of action of the muscle. This change occurs when muscle fibers are displaced as the result of contraction or when the angle between tendon and bone changes during joint rotation. Any shift in the location of the instantaneous axis of rotation will also produce a change in the length of the muscle moment arm. In the past, moment arm measurements were generally determined from measurements made on cadavers because tendon displacement and tendon force could not be measured noninvasively. However, today noninvasive imaging techniques such as magnetic resonance imaging (MRI) and ultrasound can be used to obtain measurements in different passive and active muscle states. Muscle moment arms are sometimes estimated from serial cross-sectional MRIs of muscle or tendon. The geometric center of each muscle or tendon cross-section is determined, and these centroid points are joined by a curved line. The line of action of muscle force at a given location is taken as the line tangent to the centroid line at that location. The moment arm is then the shortest distance between the axis of rotation and the line of action.

Ultrasound imaging of the point of insertion of a muscle *fascicle* into the aponeurosis has also been used to estimate the moment arms of certain muscles. The moment arm can be estimated from the ratio of change in position of the fascicle insertion point to change in joint angle. One advantage of this technique is the ability to determine how the moment arm changes as a function of joint angle and to determine whether it depends on the level of muscle contraction. However, it can be applied to only a relatively small number of superficial muscles in which muscle fascicle arrangements provide sufficiently high contrast between muscle fibers and aponeurosis to reliably measure the displacement of muscle fibers.

Muscle moment arms tend to be rather small. For example, the largest moment arm of any muscle crossing the wrist is about 2 cm. At the elbow, flexion/extension moment arms are generally less than 3 cm, although for certain joint angles the moment arm of the brachioradialis muscle can exceed 7 cm. Note that brachioradialis is somewhat unique in that its muscle fibers cross the joint rather than its tendon. In the case of brachioradialis, the moment arm increases as the muscle pulls away from the joint during contraction.

Most muscles produce moments about several axes of rotation. For example, the biceps muscle produces supination of the forearm, as well as flexion of the elbow. A muscle's moment arms are different for different axes of rotation. All joints are spanned by several muscles, each of which produces torque in a direction that depends on its line of action. In general, each muscle has separate attachments to bone and has a unique direction of action that is likely to produce torque around an axis of rotation that is not aligned with the axis of rotation of the joint. Off-axis components of torque must be balanced to control rotation about a specific axis. If the joint is a simple one-degree-of-freedom joint, then motion is mechanically constrained and any off-axis torque is balanced by reaction forces in bones and connective tissue. If the joint has more than one degree of freedom, it will generally be necessary to control the force generated by several muscles with different lines of action in order to cancel off-axis torques. If one muscle has groups of fibers oriented in multiple directions, it may be possible to change the torque contribution of that muscle around a particular axis of rotation by selectively activating different groups of muscle fibers.

Several factors that contribute to variability (noise) in motor output have been described in chapters 2 and 3. These include the dependence of muscle force on muscle length, which changes with joint angle, as well as its dependence on past activation history. In pennated muscle, the angle of pennation can also change significantly with joint angle and with the level of muscle activation (Herbert & Gandevia, 1995). This change affects the fraction of muscle force that contributes to muscle torque at the joint. In addition, the moment arm of a motor unit can vary with joint angle or with the level of muscle activation as the muscle fibers deform, displacing their line of action with respect to the joint center of rotation. These effects are essentially due to differences in initial conditions and limit the CNS's ability to use one motor command to repeatedly produce identical movements of the joint. Milner (1986a) showed that the CNS does not automatically take these factors into account, leading to predictable errors in achieving a target joint angular velocity when knowledge of results was withheld.

Muscles acting across a joint are classified as either synergistic or antagonistic. *Synergistic* muscles produce torques that cause the same rotation of a joint; *antagonistic* muscles produce torques that cause opposite rotations of the joint. Because some muscles have multiple actions at a single joint and because muscles can produce reaction torque at distant joints, pairs of muscles that act antagonistically in one task may act synergistically in another. The forces created by antagonistic muscles counteract each other, resulting in net joint torque τ, which is equal to the difference of their torques:

$$\tau = \tau_+ - \tau_- = \rho_+ \mu_+ - \rho_- \mu_- \tag{4.2}$$

where τ_+ is the joint torque created by the force μ_+ produced by muscles that rotate the joint in one direction with a moment arm ρ_+ and τ_- is the joint torque created by the force μ_- produced by muscles that rotate the joint in the opposite direction with a moment arm ρ_-.

4.3 Joint Viscoelasticity and Mechanical Impedance

Joint stiffness and damping are defined in a similar way to muscle stiffness and damping, except that they are rotational quantities; joint torque is proportional to the muscle moment arm, whereas joint stiffness and viscosity are proportional to the square of the moment arm:

$$K = \frac{d\tau}{dq} = \rho\frac{d\mu}{dq} = \rho\frac{d\mu}{d\lambda/\rho} = \rho^2\frac{d\mu}{d\lambda} = \rho^2 K_\mu \tag{4.3}$$

$$D = \frac{d\tau}{d\dot{q}} = \rho\frac{d\mu}{d\dot{q}} = \rho\frac{d\mu}{d\dot{\lambda}/\rho} = \rho^2\frac{d\mu}{d\dot{\lambda}} = \rho^2 D_\mu \tag{4.4}$$

where K is the joint stiffness, τ the joint torque, q the joint angle, μ the muscle force, ρ the moment arm, λ the muscle length, K_μ the muscle stiffness, D the joint viscosity, $\dot{q}$ the joint angular velocity, $\dot{\lambda}$ the rate of change of muscle length, and D_μ the muscle viscosity.

When a joint is displaced by a small angle, muscles crossing the joint that cause rotation in one direction will be stretched and muscles that cause rotation in the opposite direction will be shortened. The displacement increases the force of the muscles that are stretched and decreases the force of the muscles that are shortened. As a consequence, there is summation of the joint stiffness and viscosity arising from all muscles crossing the joint:

$$K = \frac{d\tau}{dq} = \frac{d(\tau_+ - \tau_-)}{dq} = \frac{\rho_+ d\mu_+}{dq} - \frac{\rho_- d\mu_-}{dq} = \frac{\rho_+ d\mu_+}{d\lambda_+/\rho_+} - \frac{\rho_- d\mu_-}{d\lambda_-/\rho_-}$$
$$= \frac{\rho_+^2 d\mu_+}{d\lambda_+} - \frac{\rho_-^2 d\mu_-}{-d\lambda_-} = \rho_+^2 K_{\mu+} + \rho_-^2 K_{\mu-} \tag{4.5}$$

$$D = \frac{d\tau}{d\dot{q}} = \frac{d(\tau_+ - \tau_-)}{d\dot{q}} = \frac{\rho_+ d\mu_+}{d\dot{q}} - \frac{\rho_- d\mu_-}{d\dot{q}} = \frac{\rho_+ d\mu_+}{d\dot{\lambda}_+/\rho_+} - \frac{\rho_- d\mu_-}{d\dot{\lambda}_-/\rho_-}$$
$$= \frac{\rho_+^2 d\mu_+}{d\dot{\lambda}_+} - \frac{\rho_-^2 d\mu_-}{-d\dot{\lambda}_-} = \rho_+^2 D_{\mu+} + \rho_-^2 D_{\mu-} \tag{4.6}$$

Because muscle stiffness and viscosity are functions of muscle force, the total joint stiffness and viscosity are also functions of muscle force. If we approximate the relation between stiffness or viscosity and muscle force as being linear, then joint stiffness and viscosity can be approximated by the following equations of muscle force:

$$K = \rho_+^2 K_{\mu+} + \rho_-^2 K_{\mu-} = \rho_+^2 (\kappa_+ + c_K \mu_+) + \rho_-^2 (\kappa_- + c_K \mu_-) \qquad (4.7)$$

$$D = \rho_+^2 D_{\mu+} + \rho_-^2 D_{\mu-} = \rho_+^2 (\delta_+ + c_D \mu_+) + \rho_-^2 (\delta_- + c_D \mu_-) \qquad (4.8)$$

where (κ_+, κ_-) and (δ_+, δ_-) are the contributions to muscle stiffness and viscosity by passive forces and c_K and c_D are coefficients relating active muscle force to muscle stiffness and damping, respectively. In a more realistic approximation, K and D would also be functions of length change and rate of length change, respectively.

An important concept in motor control is that of *mechanical impedance*, which determines the mechanical interaction between the limbs and the environment (Milner, 2009). Formally, mechanical impedance is defined as the forces that a mechanical system generates in response to imposed motions. As will be described in chapter 5, impedance is characterized by stiffness, viscosity, and inertia. In particular, inertia resists acceleration, viscosity resists velocity and stiffness resists displacement. In the case of muscle fibers, the resistance is primarily viscoelastic. However, when we consider muscles as mechanical elements connected to the skeleton, we must also consider their mass since it contributes to the inertia of a body segment. From equations (4.7) and (4.8), it is clear that the mechanical impedance of a joint will depend on the muscle viscoelasticity and the muscle moment arms. The impedance at the endpoint of a multi-joint linkage such as the arm or leg also depends on the joint angles as will become apparent in chapter 5.

Clearly, there are multiple solutions for μ_+ and μ_- in equation (4.2), with the number of possible solutions increasing with the number of muscles acting across the joint. Consequently, there is a relatively large range of possible joint impedances that can be generated for a given net joint torque (equations [4.7] and [4.8]). This range provides the nervous system with the ability to independently control the net joint torque and the joint impedance over a significant range, using *co-contraction* of antagonistic muscles, which allows considerable flexibility in selecting the dynamic behavior of the joint. In particular, we have shown that joint stiffness can be modulated over a fivefold range by means of muscle co-contraction without any change in the net joint torque (Milner et al., 1995).

4.4 Sensory Feedback Control

The stiffness and damping of a joint are not simply determined by the intrinsic mechanical behavior of muscle because the resistance to a disturbance can also be profoundly affected by *feedback* from peripheral sensory receptors. Sensory receptors that respond to changes in the mechanical state of a muscle can contribute to joint stiffness and damping by modulating the activation of α-motoneurons through neural *reflex circuits*. These feedback circuits from sensory receptors to α-motoneurons cause muscle force to change through modulation of the firing rates of the sensory neurons as a function of parameters such as change in muscle length, rate of change of length, or force.

Although feedback circuits can involve pathways that pass through the brainstem and cortex, the most rapid responses are mediated by pathways from the peripheral sensory receptors to the spinal cord and back to muscles. A spinal feedback circuit consists of specific peripheral sensory receptors, their nerve fibers and any spinal cord interneurons that receive input directly (second-order neurons) or indirectly (third or higher order interneurons) from these sensory nerve fibers, the motoneurons that receive synaptic input from these neurons, and the muscle fibers innervated by the motoneurons. A given sensory receptor is usually involved in several different spinal feedback circuits because it projects to more than one type of second-order interneuron and the output from second- and higher-order interneurons may diverge as well before reaching target motoneurons.

All motoneurons and interneurons receive *convergent* input from many sources. At any given time, their activity reflects the summation of excitatory and inhibitory postsynaptic potentials from several of these sources. The convergent inputs onto a single neuron may include sensory inputs from the periphery, descending inputs from supraspinal regions, and inputs from interneurons (figure 4.2). Convergence of descending inputs onto

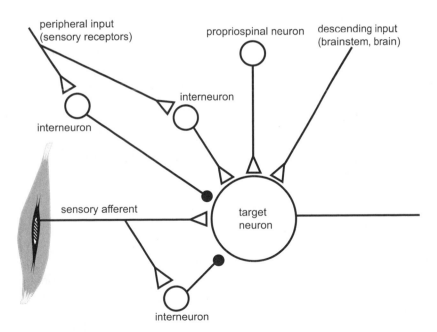

Figure 4.2
Spinal cord neurons receive converging synaptic input from many different sources, including peripheral sensory receptors, cortical neurons, brainstem neurons, and spinal interneurons. Direct inputs from sensory afferent fibers or descending fibers from higher levels tend to be excitatory, whereas inputs from interneurons can be either excitatory (open triangles) or inhibitory (filled circles).

interneurons that also receive input from peripheral sensory receptors is an important mechanism by which higher centers can enhance (amplify) or depress (attenuate) the effects of sensory input to control feedback gain. In particular, some interneurons can act as gates that control whether a peripheral input even reaches motoneurons. Gating can also be achieved directly by descending fibers acting presynaptically on the terminals of afferent fibers to produce presynaptic inhibition. The convergence of both excitatory and inhibitory inputs onto motoneurons allows for the possibility of a change in the sign of *feedback gain* from positive (excitatory) to negative (inhibitory) or vice versa (Lacquaniti, Borghese & Carrozzo, 1991).

Feedback responses are often delayed considerably (up to 100 ms or longer) with respect to the time of occurrence of the stimulus (De Serres & Milner, 1991; Capaday, Forget & Milner, 1994; Johansson & Westling, 1988) because of the time taken for action potentials to conduct along axons and transmission and activation delays that occur at synapses. In some cases, conduction pathways may be very long, ascending to sensory areas in the brainstem or brain and then descending back to the motoneurons in the spinal cord (Matthews, 1991). Feedback effects are generally distributed; that is, stimuli to the mechanoreceptors of a single muscle are functionally organized in such a way that they influence the activity of not only that muscle but also other muscles acting at the same joint and muscles acting at neighboring joints. In this way, muscle actions can be coordinated to achieve limb mechanical behavior that is appropriate for specific situations.

The most powerful feedback circuit involves the monosynaptic excitation of α-motoneurons by Ia afferent fibers, known as the *monosynaptic stretch reflex* (figure 4.3). This excitation is distributed to motoneurons of muscles that are close synergists at the same joint, as well as motoneurons of synergistic muscles at neighboring joints. The knee-jerk reflex is an example of the monosynaptic stretch reflex. When the patellar tendon is tapped, it stretches the quadriceps muscles. The muscle spindles located in the quadriceps undergo a rapid stretch that creates a burst of action potentials in Ia afferent fibers. The resulting excitatory synaptic input from Ia afferents is integrated by the quadriceps α-motoneurons, leading to their contraction and extension of the knee. In parallel with the monosynaptic excitation of α-motoneurons of the receptor bearing muscle and its synergists, activity of Ia afferents also causes *reciprocal inhibition* of α-motoneurons of antagonistic muscles. This inhibition is mediated by a class of interneurons known as Ia inhibitory interneurons. These interneurons receive convergent input from many sources, including both peripheral afferents, other interneurons, and inputs from neurons in the brain and brainstem. In the stretch reflex, Ia inhibitory interneurons mediate the reciprocal inhibition that coordinates the actions of opposing muscles. As one muscle contracts, the other relaxes.

Activation of Ib afferents, which originate from Golgi tendon organs, can lead to the inhibition of α-motoneurons of the muscle of origin. However, there is also a wide distribution of inhibitory and excitatory actions onto other motoneuron pools. The inhibitory

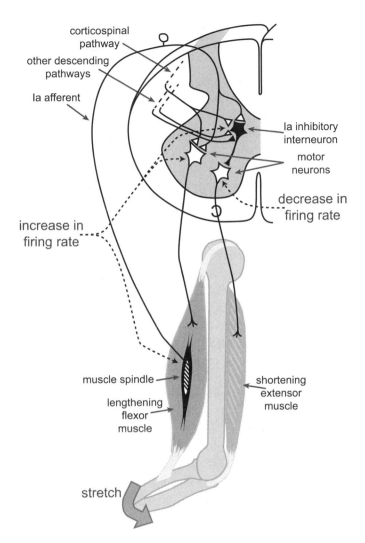

corticospinal
pathway

other descending
pathways

Ia afferent

Ia inhibitory
interneuron

motor
neurons

decrease in
firing rate

increase in
firing rate

muscle spindle

shortening
extensor
muscle

lengthening
flexor
muscle

stretch

Figure 4.3
The monosynaptic stretch reflex consists of a neural circuit originating from the Ia muscle spindle afferent of the stretched muscle and terminating on the α-motoneurons of the same and homologous muscles. It produces excitation of these motoneurons and inhibition of antagonist motoneurons through excitatory input to an interneuron referred to as the Ia inhibitory interneuron.

actions of Ib afferents are largely mediated by a class of interneurons known as Ib inhibitory interneurons. Descending input to the Ib inhibitory interneurons can act to select the pattern of distribution of the actions of Ib afferents. The Ib interneurons receive both excitatory and inhibitory inputs from descending tracts, as well as convergent input from other peripheral afferents, including monosynaptic excitation from Ia afferents and disynaptic excitation from cutaneous and joint afferents.

There are other feedback pathways, originating from cutaneous sensory afferents that are known to play important roles in motor control. They are responsible for actions such as rapidly withdrawing a limb from a painful or harmful stimulus and increasing grip force to prevent an object from slipping out of the hand. In addition, there is evidence that cutaneous sensory receptors that are sensitive to skin stretch may contribute to the feedback control of joint position (Moberg, 1983).

In robotic systems, *feedback control* is used to control a desired output variable, such as position or force. In *position control*, the difference between the actual position and the desired position is used as an error feedback signal to control the command signals to the actuators that move the robot. This type of control is called negative feedback because it is designed to reduce the error between the desired and actual position. The magnitude of the error correction depends on the feedback gain, that is, the coefficients that convert position error into command signals to the actuators. The accuracy with which a desired path can be followed depends on the magnitude of the force that is producing the deviation, the feedback gain, and the maximum torque that can be produced by the motors. The higher the feedback gain, the greater the torque that will be generated for a given error until a limit is reached. Higher feedback gains allow the robot to overcome greater force deviations. Because sensor signals always contain some noise, there is a practical limit to the maximum feedback gain that can be achieved without introducing instability. Sensor noise cannot be distinguished from true position error, so it is amplified by the feedback gain, resulting in erroneous command signals to the motors. As the feedback gain increases, these erroneous commands can become large enough to cause undesirable vibration or even instability.

In general, physical systems have mass and stiffness. These two properties endow the system with a *natural frequency* of oscillation. One of the limitations of a feedback control system that relies solely on position error is that rapid changes in position can induce oscillations at the natural frequency. Damping (velocity-dependent resistance) opposes motion and reduces the tendency for oscillation. It can also reduce the amplitude of vibration produced by sensor noise. For this reason, both position and velocity feedback are normally incorporated in feedback control. Such *proportional-derivative* (PD) controllers are frequently used in robotics applications.

Given that the firing rates of muscle spindle afferent nerve fibers are functions of both muscle length and velocity, it is conceivable that a neural feedback circuit in which muscle spindles act as the sensors could be used to control muscle length and hence the position of a joint, much like PD control of a robot. However, there are several factors that limit the effectiveness of such a control system. Most important is the delay from the time that the muscle spindle changes its firing rate in response to a change in muscle length until muscle force changes. This delay comprises the conduction time to the spinal cord, the synaptic delay, the conduction time back to the muscle, and the excitation-contraction delay. Even the fastest circuit consisting of the Ia muscle spindle afferent fibers and two synapses (at the α-motoneuron and muscle fiber) requires at least 50 ms to produce a change in force following a change in muscle length. This delay increases as the length of the feedback pathway increases, that is, by adding interneurons or distance (brainstem or cortical pathways). If the part of the body whose position is being controlled moves during the delay period the error at the time that muscle force begins to change will not match the error detected by the muscle spindles. In contrast, the feedback signal in robot controllers is almost instantaneous so the effect of delay is negligible. Once oscillation of a joint begins, muscle stretch will induce delayed activation in the stretched muscle where the delay is determined principally by the length of the feedback pathway; the higher the oscillation frequency, the later in the cycle the activation will occur. If the oscillation frequency is sufficiently high or the delay is sufficiently long, the activation can occur during the phase when the muscle is shortening. When feedback activates a muscle while it is shortening, it contributes to the motion rather than opposing it. This reinforcement of the motion creates a situation of negative impedance and will lead to oscillatory instability if the force created by feedback activation of muscles is greater than the intrinsic imped-ance created by the cross-bridges (Jacks, Prochazka & Trend, 1988; Rack, 2011), particu-larly if the body segment has little inertia.

Experiments conducted to test the ability of the monosynaptic stretch reflex as a servo-mechanism have demonstrated that its ability to compensate for unexpected perturbations to limb position is limited (Bennett, Gorassini & Prochazka, 1994). Nonetheless, it does contribute to reducing position error, although other mechanisms acting at longer delays are required to restore position after an unexpected perturbation. Other functions have been attributed to the monosynaptic stretch reflex, including compensation for nonlinear muscle properties such as yielding. When a muscle changes length, the first part of its dynamic response is a resistive force arising from its viscoelastic properties. This response is non-linear in its dependence on the direction, amplitude, and speed of the length change and may exhibit yielding during stretch. One effect of the feedback from muscle spindles is an improvement in the linearity of the dependence of the resistive force on the amplitude and direction of the length change imposed on the muscle (figure 4.4). The feedback force compensates for yield that occurs during large amplitude stretches at low and intermediate levels of muscle force in slow twitch muscle fibers (Nichols & Houk, 1976). In addition, there is a greater change in the sustained force if the muscle is held at its new length than that observed in the absence of feedback from muscle spindles. This behavior represents an enhancement of the force produced by the intrinsic muscle stiffness. During activities

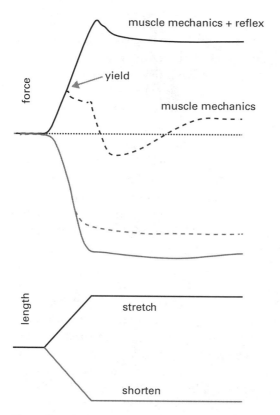

Figure 4.4
The additional muscle force created by an increase in excitatory feedback from muscle spindles during stretch (solid line) helps to compensate for yielding in slow-twitch muscle. The additional reduction in muscle force during muscle shortening (solid line) due to reduction of excitatory feedback from muscle spindles creates a more symmetric response to length changes of the same amplitude that either stretch or shorten the muscle.

such as walking or landing from a jump in which muscles are stretched by ground reaction impact forces when the foot contacts the ground, the feedback from muscle spindles would complement the force developed by voluntary activation of leg extensor muscles in resisting collapse of the leg and recovery of upright posture.

The simplest model of feedback control involving the stretch reflex assumes that the nervous system attempts to control some desired position and that any deviation from this desired position represents an error that is resisted by a change in muscle activation produced by the stretch reflex. The feedback gain is assumed to be constant, that is, the resistive force is proportional to the error. Generally, the model also incorporates a delay to account for the time taken for sensory transduction, signal conduction, synaptic transmission, muscle excitation, and so on. Although such a model is essentially identical to feedback control in a robotic system, it does not accurately represent stretch reflex

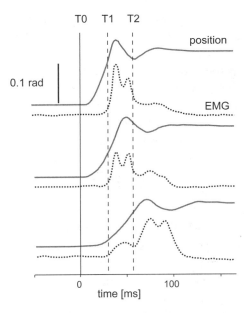

Figure 4.5
The electromyographic (EMG) response to stretch of wrist flexor muscles created by a ramp displacement of the wrist position. There is a short-latency monosynaptic activation at time T1 and a longer latency response beginning at about time T2, which likely has both transcortical and spinal components. The amplitude of the displacement is the same, but the slope of the ramp (velocity) increases going from the bottom row to the top (based on data from Lee & Tatton [1982]).

dynamics. First, the reflex response depends on the velocity of the muscle length change (Lee & Tatton, 1982). In figure 4.5, the wrist is rotated through the same angle at different velocities, showing that the change in muscle activation increases as the velocity increases. Second, the reflex response consists of components that appear at different delays and the dynamics of these components differ (Jaeger, Gottlieb & Agarwal, 1982). Third, the sustained response is relatively small compared to the transient response. This example is illustrative of the stretch reflex response when a single joint is displaced in isolation. However, most disturbances to limb position will displace several joints and invoke a coordinated resistance from many muscles resulting in dynamics, which can be considerably more complex.

There are limitations, though, to drawing conclusions about feedback control from changes in muscle activation such as those shown in figure 4.5, as the electromyographic signal is local and can be highly selective and hence a poor representation of activation throughout a muscle. Furthermore, the transformation between the electrical signal recorded from muscle and the joint torque that it creates is delayed, low-pass filtered, and nonlinear. Because changes in muscle force due to feedback from peripheral sensory receptors such as muscle spindles constitute contributions to the mechanical impedance of a joint, perturbation studies of joint position designed to quantify mechanical impedance provide a better means of characterizing feedback control than quantification of the electromyographic response. The major limitation of these studies is that they cannot reliably separate the contribution of intrinsic muscle properties to mechanical impedance from that arising from feedback circuits. Nevertheless, they provide information that is useful for deciding on the model structure for feedback control.

The most rigorous methods that have been used to quantify the mechanical impedance of human joints involve applying stochastic displacements while a person attempts to maintain a joint at a specified angle. One limitation of this approach is that the vibratory nature of the disturbance engenders co-contraction of antagonistic muscle groups, which elevates the impedance and can therefore lead to overestimation of the impedance compared to the situation in which antagonistic muscles are reciprocally activated. In addition, both the moment of inertia of the joint and its viscoelastic resistance contribute to the impedance. Therefore, a parametric model of the impedance must be assumed so that the components due to moment of inertia and viscoelasticity can be separated. This approach can introduce error if the assumptions of the model are incorrect. Nonetheless, such studies of joint impedance are still useful. The first step in characterizing the contribution of muscle activation to joint impedance is to determine joint impedance when the muscles are relaxed. When the muscles are completely relaxed, there is generally little or no activation through feedback circuits because the integrated excitatory synaptic input to α-motoneurons during the imposed displacements remains well below the threshold necessary for action potential generation. The joint impedance measured in the absence of muscle activation is referred to as the *passive impedance*. In general, passive joint stiffness and viscosity are small in the midrange of motion, but increase dramatically at the extremes of motion (figure 4.6).

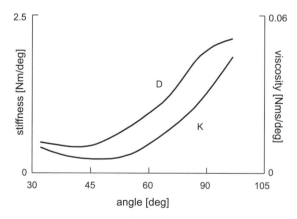

Figure 4.6
Passive ankle stiffness (K) and viscosity (D) increase as the joint is positioned closer to the end of its range of motion (based on data from Weiss, Hunter & Kearney [1988]).

When the muscles are activated, both joint stiffness and viscosity increase relative to their passive values (figure 3.9). Both tend to increase linearly as joint torque increases, that is, as muscle activation increases. Assuming that the total joint impedance is the sum of the passive impedance and that due to active processes, subtraction of the passive impedance from the total impedance yields the active component of the joint impedance. The active component appears to be proportional to joint torque over a relatively large range (Weiss, Hunter & Kearney, 1988). Furthermore, it is much less dependent on joint position than the passive impedance. However, there is no straightforward method of decomposing the active component into contributions arising from the number of attached cross-bridges (intrinsic) and changes in muscle activation (feedback). The active component of the joint stiffness depends on the amplitude of the displacement. Smaller displacements are resisted by greater stiffness than larger displacements (figure 3.9), which can be attributed to two factors. First, as the amplitude of the displacement increases, the limit of short-range stiffness is exceeded, that is, cross-bridges undergo forcible detachment. Second, the sensitivity of muscle spindles decreases with stretch amplitude, thereby effectively reducing the feedback gain as the displacement amplitude increases.

Attempts have been made to separate the joint stiffness into contributions from intrinsic and feedback sources using an alternative method based on the resistance to fast ramp and hold displacements of a joint. An estimate of the contribution of intrinsic muscle mechanics is obtained by electrically stimulating the muscles to produce a desired joint torque while a person maintains a state of complete relaxation of the limb. The feedback circuit is effectively removed because α-motoneuron dendrites and somas are disengaged by keeping the membrane potential of the soma below action potential threshold. The procedure is then repeated when the same joint torque is produced by voluntary muscle activation. In this case, feedback circuits are fully engaged. The joint torque obtained in response to displacement during electrical muscle stimulation is subtracted from that obtained during voluntary muscle activation (Leger & Milner, 2000). The difference represents the contribution of feedback circuits (figure 4.7). The results of such attempts to separate intrinsic and feedback stiffness suggest that the intrinsic stiffness increases with joint torque in the low to intermediate torque range, as expected from studies of isolated muscles. The proportion of stiffness contributed by feedback declines at high joint torques because the feedback response saturates and then declines as fewer inactive motor units are available to be recruited and fewer active motor units have the capacity to increase their firing rate (Sinkjaer et al., 1988). There are limitations to the accuracy of this approach. First, the intrinsic stiffness may be different when the joint torque is produced by electrical stimulation compared to voluntary muscle activation because fast twitch motor units tend to be activated at lower forces and slow twitch motor units at higher forces during electrical stimulation than during voluntary muscle activation. Because the stiffness of slow twitch and fast twitch motor units are different, the order in which they are activated affects stiffness values. Second, electrical stimulation of muscle is not

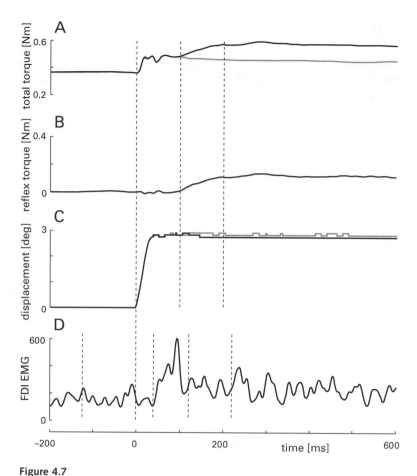

Figure 4.7
The first dorsal interosseus (FDI) muscle of the index finger is either electrically stimulated (light trace) or activated voluntarily (dark trace) to generate the same initial torque as shown in (A). The metacarpophalangeal joint of the index finger is then rapidly displaced as shown in (C) by means of a servo-controlled torque motor, stretching the FDI muscle. The difference in the change in torque produced by displacing the joint shown in (B) (dark trace minus light trace in A) represents the reflex contribution. The two vertical dashed lines preceding displacement and following displacement represent time intervals where joint position is essentially stationary. The average change in torque and displacement between these two intervals is used to calculate intrinsic and reflex stiffness. The change in muscle activation responsible for the reflex response begins about 40 ms following the onset of joint displacement as shown in (D) (adapted with permission from Leger & Milner [2000]).

painless. Therefore, it may evoke involuntary co-contraction of antagonist muscles, which would lead to overestimation of the intrinsic joint stiffness.

The inability to readily separate intrinsic and feedback contributions to joint impedance makes it difficult to propose and validate a model structure for feedback. Although it is possible to include some of the nonlinear stretch reflex dynamics in a feedback model, many models assume that the relation between sensory input (e.g., position and velocity) and motor output is linear. Feedback models may be implemented at the joint level, where the sensory inputs are joint position and joint angular velocity and the output is the joint torque, or may be implemented at the muscle level, where the sensory inputs are muscle length and the rate of change of muscle length and the output is the muscle force. We have used such a model in simulations of motor learning (Franklin et al., 2008). The more sophisticated models may incorporate feedback components with different delays. To avoid oversimplifying the dynamics, the model should take into consideration the low-pass nature of the transformation between the electrical and mechanical response of muscle, which implies that the force or torque created by sensory feedback should bear some resemblance to a low-pass filtered version of the sensory signal that produced it.

Although Golgi tendon organ afferents have been shown to produce inhibition of α-motoneurons in animal studies by means of Ib inhibitory interneurons, there is no convincing evidence that this circuit plays a significant role in the feedback control of human movement. On the other hand, studies in which the leg is unexpectedly loaded or unloaded suggest that a feedback circuit involving excitation of α-motoneurons may be engaged during normal gait (Grey et al., 2007). This circuit is an example of positive feedback. Such a circuit does not act to correct force errors; rather, it acts to balance loads. Assuming that this circuit is active during normal unperturbed gait, unloading the limb would have the effect of reducing Golgi tendon organ output, which would reduce the force supporting the body, whereas loading the limb would increase the supporting force. Thus, the feedback circuit would compensate for unexpected disturbing forces. This principle has been applied experimentally in a simple simulated feedback loop for load compensation for the wrist and ankle joints (Prochazka, Gillard & Bennett, 1997).

4.5 Voluntary Movement

The motion of a body segment about a joint is controlled in a reciprocal manner by antagonistic groups of muscles; that is, one muscle group accelerates the joint in one direction and the other group accelerates it in the opposite direction. A movement initiated from rest requires the application of torque to the joint, which generally originates from contraction of muscles. The muscles that produce acceleration in the direction of motion are referred to as *agonists* of the movement. To slow down (decelerate) the movement, torque must be applied in the opposite direction. The muscles that produce deceleration are referred to as *antagonists* of the movement (figure 4.8). Once a movement is under way,

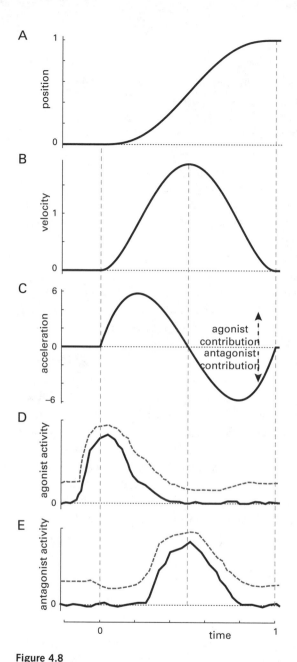

Figure 4.8
A rapid movement of a limb segment about a single joint between two positions, represented in terms of joint angular motion in (A), is initiated by a burst of muscle activity in the agonist muscles starting prior to movement (D). This produces a torque impulse, which accelerates the joint (C), during which angular velocity increases. Shortly after the joint begins accelerating, the antagonist muscles are activated to decelerate the joint (E). Frequently, the timing, amplitude, and duration of the muscle activity lead to roughly symmetric acceleration and deceleration profiles, which produces a symmetric, bell-shaped velocity profile (B). The dashed lines in (D and E) illustrate that multiple activation patterns can produce the same kinematics. In this case, there is coactivation of the agonist and antagonist muscles throughout the movement.

it can be stopped only by applying a torque impulse that is equal and opposite to the angular momentum of the moving limb segment, that is,

$$\int_{t=0}^{t=T} \tau \, dt = -I\dot{q} \tag{4.9}$$

where τ is the joint torque, I is the moment of inertia of the limb segment, and $\dot{q}$ is the angular velocity of the joint. The decelerating torque may originate from antagonist muscles and/or from external force applied to the limb segment.

Although many studies of human movement suggest that the acceleration and deceleration phases are relatively symmetric, resulting in a movement with a bell-shaped velocity profile, movements are not constrained to be symmetric (Brown & Cooke, 1990). The velocity profile is determined by the torque profile, which generates the movement. The torque profile arising from activation of agonist muscles will roughly reflect the muscle force profile, which will tend to be asymmetric, because muscles contract more quickly than they relax. When the agonist torque is applied to the skeleton, it moves a limb segment against the resistance created by its inertia and the viscoelastic properties of the tissues that change length during the movement, that is, against the mechanical impedance. Assuming that the impedance can be approximated by linear second-order mechanics, this torque will produce a smooth movement that takes less time to accelerate than to decelerate. Thus, a movement controlled purely by activation of agonist muscles would be expected to have an asymmetric velocity profile. However, by activating antagonist muscles at an appropriate delay with respect to the agonist activation, the duration of the acceleration and deceleration phases can be equalized, creating a symmetric velocity profile. By modifying the delay between agonist and antagonist activation and by modulating the duration and amplitude of the bursts of muscle activation, velocity profiles can be created with a range of asymmetry (figure 4.9). By concatenating a series of bursts of agonist and antagonist muscle activation at slight delays, for example 50–100 ms, a highly asymmetric velocity profile can be achieved that is characteristic of movements in which the final position must be controlled with a high degree of accuracy (Milner, 1992). Such movement patterns are also characteristic of deliberately slow movements (figure 4.10).

The simplest mechanical model of a limb segment is that of a mass with a moment of inertia, which gives the following equation of motion:

$$\ddot{q} = \frac{\tau}{I} \tag{4.10}$$

The angular acceleration of the joint is proportional to the torque applied to the joint. However, the peak angular velocity of the joint does not simply reflect the torque amplitude because velocity is the integral of the acceleration. Therefore, peak velocity is also a function of the duration of the torque. This relation is clear from the relation between torque

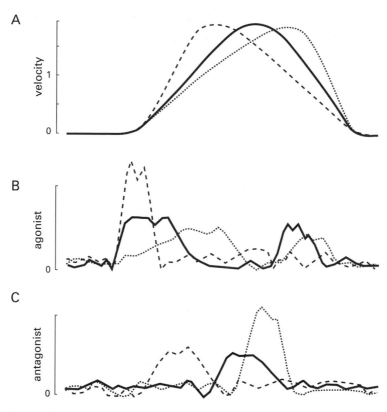

Figure 4.9
The skew of the velocity profile during rapid movements of a single joint can be controlled by varying the timing and amplitude of bursts of agonist and antagonist muscle activity. By shifting the antagonist burst to occur later in the movement and increasing its amplitude, the skew of the velocity profile can be shifted from left (dashed line) to right (dotted line). The patterns of agonist/antagonist muscle activation are shown for left skew (dashed line), no skew (solid line), and right skew (dotted line) (based on data from Brown & Cooke [1990]).

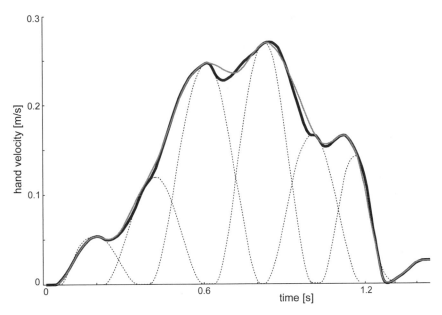

Figure 4.10
The velocity profile of a movement that is deliberately made at about a third of its normal speed appears to be comprised of a series of overlapping submovements occurring at relatively regular intervals of about 200 ms. The dotted traces represent a decomposition of the movement based on a model velocity profile for submovements that can be scaled in amplitude and duration. The lighter solid line represents the actual velocity profile and the heavier solid line represents the reconstruction based on linear summation of the submovements (adapted with permission from Milner [1992]).

impulse and angular momentum; the longer the interval during which torque is applied, the greater the change in angular momentum which it produces, that is, the higher the angular velocity. High acceleration is important in movements in which high velocity must be achieved over a brief interval. High acceleration can be achieved by means of a large burst of agonist muscle activation, that is, rapid activation of many motor units at high firing rates. However, such activation presents a problem if the amplitude of the movement is small because muscle force does not instantaneously relax. Therefore, rapid small amplitude movements would tend to overshoot the intended final position unless a portion of the torque impulse created by activation of agonist muscles is counteracted. To limit the extent of rapid, small amplitude movements, antagonist muscles must be activated shortly after agonist muscles to create a torque impulse in the opposite direction (Ghez & Gordon, 1987; Hoffman & Strick, 1990). Thus, antagonist muscle activity may act both to reduce angular momentum and to cancel a portion of the torque impulse generated by agonist muscles. One effect of this overlap of agonist and antagonist muscle activation is to increase the mechanical impedance of the joint. Co-contraction of agonist and antagonist muscles will increase both joint stiffness and viscosity, increasing stability around the final position.

Increasing the speed of a movement or increasing the load that must be moved requires increased activation of agonist muscles and also increases the mechanical impedance of the joint. The effect of this increased mechanical impedance is evident when an unexpected external perturbation, such as a torque pulse, displaces the joint from its normal trajectory (Milner, 1993). Once the perturbation has ended, the elastic component of the impedance acts to move the joint back toward the original trajectory. The faster the movement or the greater the load, the higher the stiffness, which results in less displacement and faster return to the planned (unperturbed) trajectory (figure 4.11).

The dynamics of the movements illustrated in figure 4.11 suggest that it is more accurate to model the motion of a limb segment as a linear second-order system than as the motion of a simple inertia (equation [4.10]), that is,

$$\tau(t) = I\ddot{q}(t) + D(\dot{q}_u(t) - \dot{q}(t)) + K(q_u(t) - q(t)) \tag{4.11}$$

where I is the limb inertia, $q_u(t)$ is the desired trajectory, and D and K are joint viscosity and stiffness, respectively. The elastic response in this model is similar to a linear PD controller used in robotics. Note that D and K are not constant but vary with the amount of muscle activation. Furthermore, torque and impedance are adapted to the dynamics of the activity. For example, it possible to create interactions with negative impedance that assists imposed motion. When subjects interact with negative viscosity, they cannot initially stabilize the limb around the target; that is, the limb oscillates around the target (figure 4.12). Feedback from muscle spindles which are stretched during alternating phases of the oscillation, create oscillating EMG patterns in antagonistic muscles. However, after repeating the movement many times, subjects learn to reduce the oscillation, indicating an adaptation of the limb viscoelastic properties to stabilize the limb (Milner & Cloutier, 1993).

4.6   Summary

In this chapter, we modeled the mechanics of single-joint movements and discussed issues related to feedback control. We described the kinematics and dynamics of human joints and presented the equations that relate muscle and joint dynamics. An important concept introduced is the reciprocal nature of the control of joint motion and torque. Muscles are organized in antagonistic pairs that control the direction of motion and torque at the joint in a reciprocal manner. This organization confers the ability to independently control the joint torque and the joint mechanical impedance. We described sensory feedback circuits involved in the control of human movement and compared the control of a human joint with servo-control of a single-joint robot. Although it is difficult to estimate feedback gains in human sensory feedback circuits, it is clear from experiments that the CNS cannot

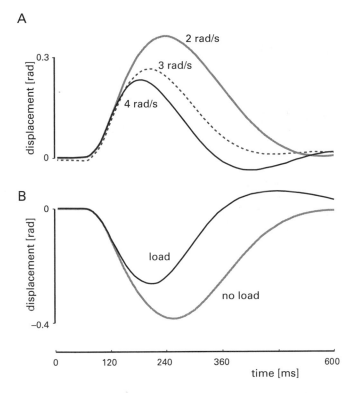

Figure 4.11
A 50 ms 5 Nm torque pulse applied by a torque motor during voluntary rotation of the elbow displaces the elbow from its usual unperturbed trajectory. The displacements shown in (A) and (B) were obtained by subtracting the unperturbed trajectory from the perturbed trajectory. As shown in (A), the displacement produced by the perturbation decreases as the peak velocity of elbow flexion increases from 2 rad/s to 4 rad/s. In addition, the duration of the displacement decreases, which is indicative of higher joint stiffness. The displacements in (B) were obtained with a torque pulse in the opposite direction to (A) under conditions in which the subject was rotating the elbow at the same peak velocity either without a load (no load) or against a viscous load (load) created by the torque motor. It is clear that the displacement also decreases when the movement is made at the same velocity but against a resistive load compared to no load (adapted with permission from Milner [1993]).

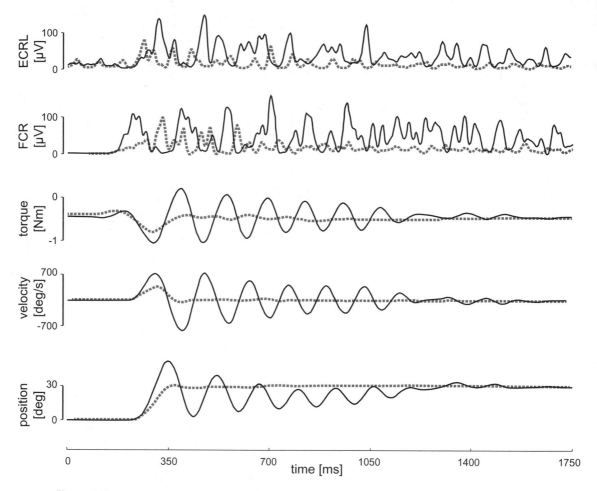

Figure 4.12
When a subject first interacts with negative viscosity during a rapid wrist movement, the wrist oscillates around the target position (dark trace). An oscillating EMG pattern in flexor (FCR) and extensor (ECRL) muscles is observed, which appears to be reinforced by stretch reflex feedback. However, after training the subject is able to reduce the oscillation (dashed line); this reduction is accompanied by a reduction in the EMG of both the flexor and extensor muscles (adapted with permission from Milner & Cloutier [1993]).

achieve the performance of a robot servo-controller. The sensory feedback gains are limited by neural noise and transmission delays. Nevertheless, there is considerable versatility in the way that sensory feedback can be used to modulate the mechanics of a single joint. The mechanical behavior of a single joint in response to unexpected perturbations suggests that a second-order linear model adequately captures the most salient features of the dynamics during voluntary movement. The ability to adaptively control the viscoelastic parameters of the joint mechanics is illustrated during adaptation to voluntary movement in the presence of unstable loads.

5 Multijoint Multimuscle Kinematics and Impedance

Whether a muscle acts as an agonist, antagonist, or stabilizer depends on kinematic factors such as joint angles, link lengths, moment arms, and force direction. Many muscles or their tendons span more than one joint (figure 5.1). For example, the biceps brachii and triceps longus in the arm as well as the hamstrings in the leg are biarticular muscles. The tendons of some finger flexors and extensors cross up to four joints, so most of the muscle mass is removed from the hand, reducing the inertia and allowing rapid response.

In multiarticular movements, the coupling between the joints will often produce undesired torques that must be counterbalanced by muscle action (Zajac, 1993; figure 5.2). Thus, most movements require the coordination of muscles acting at several joints. Some muscles act as movers, some as stabilizers, and others as brakes.

The limbs and muscles form a highly redundant system with nonlinear and coupled dynamics. It is not a straightforward matter to predict the activation pattern needed to produce a desired movement because muscles act on multiple limb segments and frequently affect the motion about more than one axis of rotation. However, kinematics, dynamics, and control provide powerful tools to describe the trajectories, endpoint force, and muscles activation, which are described in this chapter.

5.1 Kinematic Description

Most frequently, humans interact with their environment by making contact with their hands or feet. Their limbs are open kinematic chains consisting of a series of limb segments and joints (figure 5.1). It is often desirable to simplify the representation of the joints which comprise a limb to avoid the complexity that would arise if an attempt was made to be anatomically correct.

Typically we identify the *degrees of freedom* (DOF) of a limb by selecting a set of independent displacements and/or rotations that can specify completely the change in position and orientation of the limb. This gives rise to the *joint space* $\left\{ \mathbf{q} = (q_1, q_2, \ldots, q_N)^T \right\}$. For example, the elbow joint has one DOF. The wrist has several DOF, but we can restrict

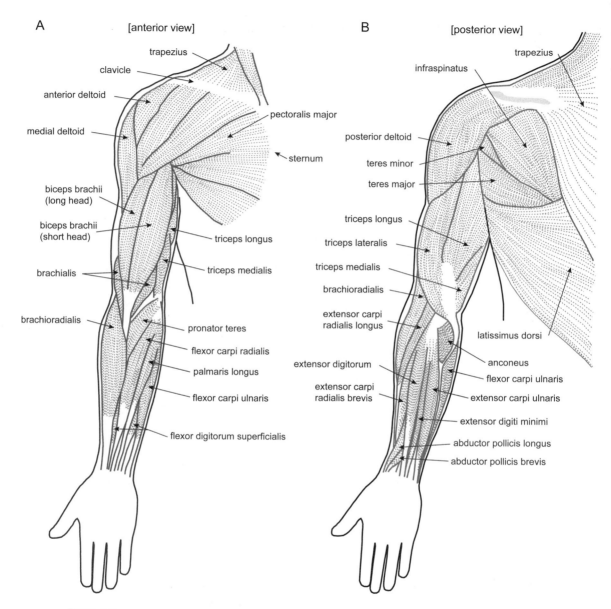

Figure 5.1
Superficial musculature of the arm. (A) Anterior view of the surface muscles of the arm. These muscles flex the joints that they cross, that is, shoulder, elbow, or wrist (note that the biceps crosses the shoulder and elbow joints). (B) Posterior view of the superficial muscles of the arm. These muscles extend the joints that they cross (note that the long head of the triceps crosses both shoulder and elbow joints). Many of the muscles shown in the forearm (both in the anterior and posterior view) are primarily used for finger flexion or extension. However, because their tendons cross the wrist, they also produce torque at the wrist joint. The figure does not show deep muscles, some of which have similar functions.

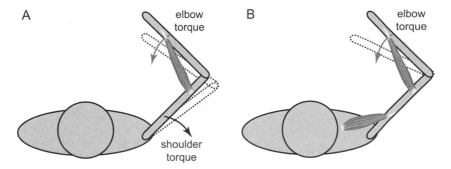

Figure 5.2
Due to the coupling between joints, muscles produce torques not only at the joints they cross but also at remote joints. (A) Elbow flexion produced by a single-joint elbow muscle is accompanied by shoulder extension torque. (B) To move only the elbow, it is thus necessary to produce simultaneously shoulder flexion torque in order to counteract the interaction torques.

the analysis to a single degree of freedom such as wrist flexion/extension and consider it as a 1DOF system, as was done in chapter 4.

The endpoint location of a limb, at the hand or the foot, can be described as a *pose* $(\mathbf{x},\boldsymbol{\theta})$ in a *task space*, where $\mathbf{x} = (x_1, x_2, x_3)^T$ is its *position* and $\boldsymbol{\theta} = (\theta_1, \theta_2, \theta_3)^T$ its *orientation*. This *extrinsic* Cartesian representation is related to visual space via visuomotor coordination. The orientation can be represented in different ways (Siciliano et al., 2009) such as quaternions or rotation matrices and must consider the fact that rotations do not commute, though this topic is beyond the scope of this book.

Besides the *extrinsic* task or hand space, and the *intrinsic* joint space, there exists another physiologically relevant intrinsic representation, namely the *muscle space* $\{\boldsymbol{\lambda} = (\lambda_1, \ldots, \lambda_M)^T\}$. The task space has maximally six DOF but fewer DOF are required to describe constrained tasks. For example moving a planar object on a table involves three DOF: two Cartesian DOF to describe position and one DOF to describe orientation. Joint space can have more DOF than the task space, depending on the number of body segments involved, and the muscle space can have as many DOF as the number of skeletal muscles that can move those segments.

5.2 Planar Arm Motion

The arm has at least seven DOF in joint space: three rotations for the shoulder, one for the elbow, and three for the wrist, including forearm rotation. However, for planar movements this model can be simplified to two DOF, as shown in figure 5.3. This 2-joint 6-muscle model of planar arm movements is very useful, as it is simple, and thus mathematically tractable and intuitive to interpret, while still possessing fundamental features

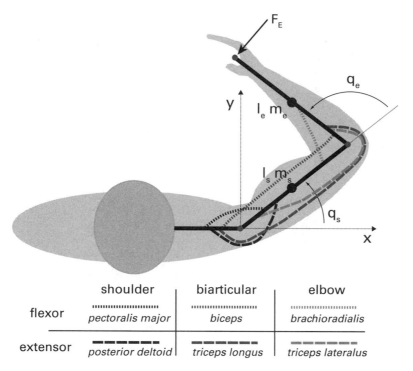

Figure 5.3
Graphical representation of a 2-joint 6-muscle model to analyze horizontal arm movements. The moment arms vary with the muscle and joint and also with the configuration.

of more complex musculoskeletal models, such as nonlinear dynamics and actuation via a redundant system of muscles.

In this model, the hand position $\mathbf{x} = (x, y)^T$ relative to the shoulder can be simply computed from the joint angles $\mathbf{q} = (q_s, q_e)^T$ and limb segment lengths $\mathbf{l} = (l_s, l_e)^T$:

$$x = l_s \cos(q_s) + l_e \cos(q_s + q_e)$$
$$y = l_s \sin(q_s) + l_e \sin(q_s + q_e)$$

(5.1)

5.3   Direct and Inverse Kinematics

The endpoint position of a limb (or of any open kinematic chain) can be computed iteratively from the joint angles by propagating the position and orientation of successive joints. Algorithms have been developed in robotics textbooks such as Spong, Hutchinson & Vidyasagar (2006) and implemented in software packages (Featherstone, 2008). These algorithms enable us to compute the transformation from joint to hand (Cartesian) space or *direct kinematics*. Similarly, the muscle lengths can be computed from the joint angles (and the moment arms).

However, what the brain must determine is how to move the joints in order to produce a desired movement or action in task space. This *inverse kinematics transformation* sometimes has no solution—for example, it is not possible to realize a hand orientation requiring a wrist angle outside the joint limit—or, conversely, has multiple solutions—while holding a glass at a fixed position one can move the elbow up and down.

As an example, we compute the inverse kinematics of the 2-joint planar model of figure 5.3, using the variables of figure 5.4A:

$$
\begin{aligned}
& x^2 + y^2 = l_s^2 + l_e^2 - 2l_s l_e \cos(\pi - q_e) \\
& \rightarrow q_e = \arccos\left(\frac{x^2 + y^2 - l_s^2 - l_e^2}{2 l_s l_e}\right) \\
& q_s = \alpha - \beta, \quad \alpha = \arctan 2(y, x) \\
& l_e^2 = x^2 + y^2 + l_s^2 - 2l_s \sqrt{x^2 + y^2}\cos\beta \\
& \beta = \arccos\left(\frac{x^2 + y^2 + l_s^2 - l_e^2}{2 l_s \sqrt{x^2 + y^2}}\right)
\end{aligned}
\tag{5.2}
$$

It is evident that even for this simple structure, the inverse kinematics are relatively complex.

5.4 Differential Kinematics and Force Relationships

An approach to circumventing the analytical and numerical difficulties in computing the direct and inverse kinematics consists of deriving the kinematics as linear relationships between the representation of velocity in different coordinate systems:

$$
\frac{dx_i}{dt} = \sum_{j=1}^{N} \frac{\partial x_i}{\partial q_j}\frac{\partial q_j}{\partial t} \quad \forall i, \text{ that is, } \dot{\mathbf{x}} = \mathbf{J}(\mathbf{q})\,\dot{\mathbf{q}}
\tag{5.3}
$$

where $\forall$ denotes *for all* and

$$
\mathbf{J}(\mathbf{q}) = \left(\frac{\partial x_i}{\partial q_j}\right)
\tag{5.4}
$$

is the *Jacobian* used to transform the velocity from the q to the x coordinate system. As an example, the Jacobian for the 2-joint model of equation (5.1) is

$$
\mathbf{J}(\mathbf{q}) = \begin{bmatrix} -l_s s_s - l_e s_{se} & -l_e s_{se} \\ l_s c_s + l_e c_{se} & l_e c_{se} \end{bmatrix}
\tag{5.5}
$$

$s_s = \sin(q_s), \; s_{se} = \sin(q_s + q_e), \; c_s = \cos(q_s), \; c_{se} = \cos(q_s + q_e)$

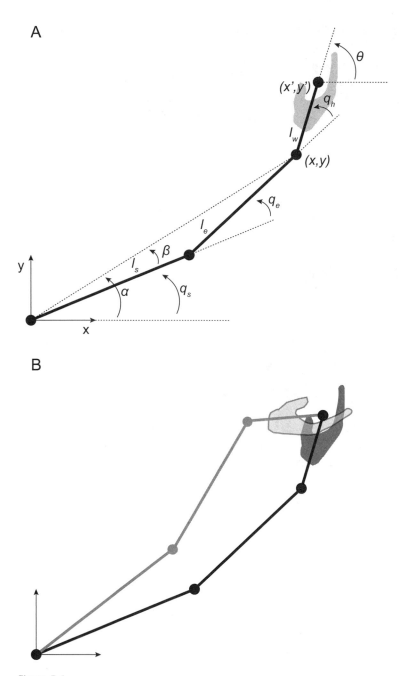

Figure 5.4
Three-joint model of planar arm movements with shoulder, elbow, and wrist flexion/extension. (A) Diagram with variables definition. (B) Kinematic redundancy: how the joint angles can be combined in various ways to grasp an object such as a ping-pong ball on a table.

Generalizing the relationship of equation (5.3), differential kinematic relationships can also be derived between the joint and muscle coordinates:

$$\dot{\lambda} = \mathbf{J}_\mu(\rho)\dot{\mathbf{q}} \tag{5.6}$$

where the Jacobian

$$\mathbf{J}_\mu(\rho) = \left(\frac{\partial \lambda_i}{\partial q_j}\right) \tag{5.7}$$

representing the differential relationship between $\mathbf{q}$ and λ coordinates is a matrix with the muscle moment arms ρ as coefficients.

For example, for a 2-joint muscle model of the arm with antagonistic pairs of uniarticular and biarticular muscles

$$\mathbf{J}_\mu(\rho) = \begin{bmatrix} \rho_{s+} & 0 \\ -\rho_{s-} & 0 \\ \rho_{bs+} & \rho_{be+} \\ -\rho_{bs-} & -\rho_{be-} \\ 0 & \rho_{e+} \\ 0 & -\rho_{e-} \end{bmatrix} \tag{5.8}$$

where $\rho = (\rho_{s+}, \rho_{s-}, \rho_{bs+}, \rho_{bs-}, \rho_{be+}, \rho_{be-}, \rho_{e+}, \rho_{e-})$ are grouped as moment arm pairs of uniarticular shoulder and biarticular and uniarticular elbow muscles, respectively.

The Jacobian relates not only the velocities in two coordinate systems but also the endpoint force, the joint torque vector and the vector of muscle tensions (figure 5.3). The relation between coordinate systems can be established from the power

$$(\mathbf{F}, \dot{\mathbf{x}}) = \mathbf{F}^T\dot{\mathbf{x}} = \sum_{i=1}^{n} F_i\dot{x}_i \tag{5.9}$$

(where the scalar product $(\cdot,\cdot)$ is defined as above), which is independent of any coordinate system:

$$(\tau, \dot{\mathbf{q}}) = (\mathbf{F}, \dot{\mathbf{x}}) = (\mathbf{F}, \mathbf{J}\dot{\mathbf{q}}) = \mathbf{F}^T\mathbf{J}\dot{\mathbf{q}} = \left(\mathbf{J}^T\mathbf{F}\right)^T\dot{\mathbf{q}} = \left(\mathbf{J}^T\mathbf{F}, \dot{\mathbf{q}}\right) \quad \forall \dot{\mathbf{q}} \tag{5.10}$$

yielding

$$\tau = \mathbf{J}(\mathbf{q})^T\mathbf{F} \tag{5.11}$$

This relation is very useful in computing the torque that must be exerted at each joint in order to produce a required force at the hand. The analogous relationship between the muscle tension vector μ and torque vector τ can be derived as

$$\tau = \mathbf{J}_\mu(\rho)^T\mu \tag{5.12}$$

5.5 Mechanical Impedance

A major role of the motor system is to interact with the environment. This interaction is two-way, as expressed in Newton's third law: action and reaction are equal, opposite, and simultaneous. Focusing on the human side, the mechanical interaction with the environment modulated by the neural input is characterized by the resistance to imposed motion or mechanical impedance (Hogan, 1985), as defined in chapter 4. A large component of mechanical impedance resisting human movement arises from the viscoelastic tissue properties. As discussed in chapter 4 for one-joint motion (figure 4.11), when the arm is slightly perturbed during movement it tends to return to its undisturbed trajectory (Milner, 1993; Won & Hogan, 1995; Gomi & Kawato, 1997), as if it is connected to a spring moving along a planned trajectory. This spring-like property stems mainly from muscle elasticity and the stretch reflex, which produce a restoring force toward the undisturbed trajectory.

Although voluntary movements are limited by the slow response of neural processes, impedance responds instantaneously. It is thus fundamental to tasks involving fast dynamic changes such as during impact with the ground in walking/running. The role of adaptive impedance control in stability is discussed later in this chapter as well as in chapter 8.

It is the mechanical behavior at the endpoint of the limb that is generally most important because most interactions take place there. If one considers that the neural control of the human arm (expressed as force $\mathbf{F}$) depends on position $\mathbf{x}$, velocity $\dot{\mathbf{x}}$, acceleration $\ddot{\mathbf{x}}$, and muscle activation $\mathbf{u}(\mathbf{x}, \dot{\mathbf{x}})$:

$$\mathbf{F} = \mathbf{F}(\mathbf{x}, \dot{\mathbf{x}}, \ddot{\mathbf{x}}, \mathbf{u}) \tag{5.13}$$

then impedance also depends on these variables. Therefore, in principle, impedance can be modulated both by changing the posture, which yields an immediate response to perturbation, and by controlling the neural input through reflexes and voluntary movements, which involves longer delays (Hogan, 1990).

In general, impedance is a nonlinear function of the kinematic variables. Linearization of the endpoint force as a function of the position, velocity, acceleration, and muscle activation yields

$$\Delta \mathbf{F}(\mathbf{x}, \dot{\mathbf{x}}, \ddot{\mathbf{x}}, \mathbf{u}) = \mathbf{K}_x \Delta \mathbf{x} + \mathbf{D}_x \Delta \dot{\mathbf{x}} + \mathbf{I}_x \Delta \ddot{\mathbf{x}} \tag{5.14}$$

where we define

$$\mathbf{K}_x \equiv \left(-\frac{\partial F_i}{\partial x_j} - \sum_k \frac{\partial F_i}{\partial u_k} \frac{\partial u_k}{\partial x_j} \right) \text{ as the } \textit{stiffness}$$

$$\mathbf{D}_x \equiv \left(-\frac{\partial F_i}{\partial \dot{x}_j} - \sum_k \frac{\partial F_i}{\partial u_k} \frac{\partial u_k}{\partial \dot{x}_j} \right) \text{ as the } \textit{viscosity, and}$$

$$\mathbf{I}_x \equiv \left(-\frac{\partial F_i}{\partial \ddot{x}_j} \right) \text{ as the } \textit{inertia.}$$

Note that these definitions are adapted from the usual definition of passive physical properties, which does not include an activation-dependent component. We can consider the mechanical impedance of biological systems to be composed of stiffness, viscosity, and inertia defined in the linearization of equation (5.14).

Current methods for measuring stiffness and viscosity of biological systems cannot isolate the activation-dependent components (u terms), that is, reflex feedback, from muscle intrinsic viscoelastic properties. For example, the method of Burdet et al. (2000) to estimate endpoint stiffness during movement displaces the hand by a constant distance relative to a prediction of the unperturbed trajectory and measures the restoring force. The perturbation consists of a 100 ms smooth ramp displacement of 8 mm that is then held constant for 100 ms, during which the restoring force is measured (figure 5.5). The restoring force includes contributions from both intrinsic muscle elasticity and reflexes corresponding to the definitions of $\mathbf{K}_x$ and $\mathbf{D}_x$ in equation (5.14).

The Laplace transform (defined, for example, in Riley, Hobson & Bence [2006]) of equation (5.14) yields

$$L[\Delta\mathbf{F}](s) = \frac{\mathbf{K}_x}{s}\Delta\mathbf{V} + \mathbf{D}_x\Delta\mathbf{V} + \mathbf{I}_x s\Delta\mathbf{V} = \left(\frac{\mathbf{K}_x}{s} + \mathbf{D}_x + s\mathbf{I}_x\right)\Delta\mathbf{V}(s) \qquad (5.15)$$

where $\mathbf{V}$ is the Laplace transform of velocity $\mathbf{v} \equiv \dot{\mathbf{x}}$. Assuming that impedance is represented by the linear equation (5.14), it can be represented as the transfer function

$$\left(\frac{L[\Delta F_i]}{L[\Delta v_j]}(s)\right) = \frac{\mathbf{K}_x}{s} + \mathbf{D}_x + s\mathbf{I}_x \qquad (5.16)$$

Measuring the response to sinusoidal signals (or to controlled stochastic perturbations) yields a Bode plot from which impedance can be computed and stiffness, damping, and inertia estimated using a parametric model (Trumbower et al., 2009).

Due to nonlinear and history-dependent muscle mechanical properties and reflex feedback, described in chapters 3 and 4 (e.g., figure 3.9), quantification of the impedance depends on the type of perturbation used for identification. In particular, it was shown that the magnitude of stiffness (identified under static conditions by displacing the hand as described previously) decreases with the amplitude of the displacement (Shadmehr, Mussa-Ivaldi & Bizzi, 1993), which may be attributed to the short-range stiffness of muscles described in section 3.5 of chapter 3. Different types of perturbations are subject to different confounding effects. Predictable perturbations such as sinusoidal displacements can be problematic at low frequencies, as subjects may voluntarily modulate their resistance at the perturbation frequency, whereas stochastic perturbations can underestimate impedance by suppressing reflex responses. Because of the many nonlinear properties of the neuromuscular system, it is important to carefully consider the experimental conditions when comparing results obtained using linear models of impedance.

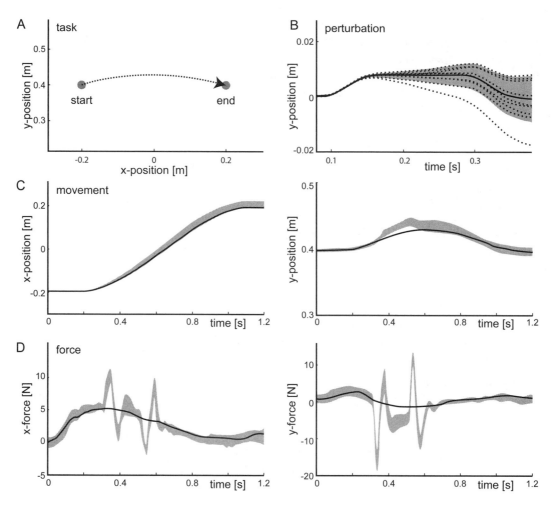

Figure 5.5
Stiffness estimation by displacing the hand during movement (Burdet et al., 2000). The mean displacement of ten trials and the deviation around a prediction (B) of the undisturbed trajectory for movement in the *x*-direction (A). (C) and (D) show the trajectory and the force elicited by the displacement in the *y*-direction.

5.6 Kinematic Transformations

In section 5.4, we discussed how the velocity and force are transformed between task, joint, and muscle coordinates. Impedance measured at the endpoint can also be transformed to the joints. In particular, joint stiffness is represented as

$$\mathbf{K} \equiv \left(\frac{\partial \tau_i}{\partial q_j} \right) = \left(\frac{\partial \left(\mathbf{J}^T \mathbf{F} \right)_i}{q_j} \right) = \left(\sum_k \frac{\partial \left(\mathbf{J}^T \right)_{ik}}{\partial q_j} F_k \right) + \mathbf{J}^T \left(\frac{\partial F_i}{\partial q_j} \right)$$

$$= \left(\sum_k \frac{\partial \left(\mathbf{J}^T \right)_{ik}}{\partial q_j} F_k \right) + \mathbf{J}^T \left(\sum_k \frac{\partial F_i}{\partial x_k} \frac{\partial x_k}{\partial q_j} \right)$$

which yields

$$\mathbf{K} = \frac{d\mathbf{J}^T}{d\mathbf{q}} \mathbf{F} + \mathbf{J}^T \mathbf{K}_x \mathbf{J} \tag{5.17}$$

where the *transpose* of the Jacobian is defined as $\left(\mathbf{J}^T \right)_{ij} \equiv J_{ji}$. The first term on the right of equation (5.17) arises from the dependence of the Jacobian on the position and varies with the external (or applied) force. Note that in the case of rotations, the transformations are more complex; see Zefran & Kumar, 2002. The relationships

$$\mathbf{D} = \mathbf{J}^T \mathbf{D}_x \mathbf{J} \tag{5.18}$$

and

$$\mathbf{I} = \mathbf{J}^T \mathbf{I}_x \mathbf{J}$$

can be derived similarly. For example,

$$\mathbf{D} \equiv \left(\frac{\partial \tau_i}{\partial \dot{q}_j} \right) = \left(\frac{\partial \left(\mathbf{J}^T \mathbf{F} \right)_i}{\partial \dot{q}_j} \right) = \left(\sum_k \frac{\partial \left(\mathbf{J}^T \right)_{ik}}{\partial \dot{q}_j} F_k \right) + \mathbf{J}^T \left(\frac{\partial F_i}{\partial \dot{q}_j} \right) = \mathbf{0} + \mathbf{J}^T \left(\sum_k \frac{\partial F_i}{\partial \dot{x}_k} \frac{\partial \dot{x}_k}{\partial \dot{q}_j} \right)$$

$$\frac{\partial \dot{x}_k}{\partial \dot{q}_j} = \sum_l \frac{\partial \left(J_{kl} \dot{q}_l \right)}{\partial \dot{q}_j} = \sum_l \frac{\partial J_{kl}}{\partial \dot{q}_j} \dot{q}_l + \sum_l J_{kl} \frac{\partial \dot{q}_l}{\partial \dot{q}_j} = 0 + J_{kj} \tag{5.19}$$

Transformation of impedance components between joint and muscle coordinates can be represented by similar relations:

$$\mathbf{K} = \frac{d\mathbf{J}_\mu^T}{d\mathbf{q}} \boldsymbol{\mu} + \mathbf{J}_\mu^T \mathbf{K}_\mu \mathbf{J}_\mu \tag{5.20}$$

$$\mathbf{D} = \mathbf{J}_\mu^T \mathbf{D}_\mu \mathbf{J}_\mu$$

These relations reveal the role of muscle arrangement and body segment geometry on the dynamics of the body when interacting with the environment.

As an example, let us examine the influence of muscle stiffness and monoarticular reflexes on joint stiffness using the two-joint arm model. We consider the case of no external force and assume unit moment arms and no biarticular reflexes, so that the muscle stiffness matrix is diagonal. The joint stiffness matrix can then be computed as

$$
\mathbf{K} = \begin{bmatrix} 1 & -1 & 1 & -1 & 0 & 0 \\ 0 & 0 & 1 & -1 & 1 & -1 \end{bmatrix} \begin{bmatrix} K_{s+} & 0 & 0 & 0 & 0 & 0 \\ 0 & K_{s-} & 0 & 0 & 0 & 0 \\ 0 & 0 & K_{b+} & 0 & 0 & 0 \\ 0 & 0 & 0 & K_{b-} & 0 & 0 \\ 0 & 0 & 0 & 0 & K_{e+} & 0 \\ 0 & 0 & 0 & 0 & 0 & K_{e-} \end{bmatrix} \begin{bmatrix} 1 & 0 \\ -1 & 0 \\ 1 & 1 \\ -1 & -1 \\ 0 & 1 \\ 0 & -1 \end{bmatrix}
$$

$$
= \begin{bmatrix} K_{s+} + K_{s-} + K_{b+} + K_{b-} & K_{b+} + K_{b-} \\ K_{b+} + K_{b-} & K_{e+} + K_{e-} + K_{b+} + K_{b-} \end{bmatrix} \equiv \begin{bmatrix} K_s + K_b & K_b \\ K_b & K_e + K_b \end{bmatrix}
$$

(5.21)

where $K_s = K_{s+} + K_{s-}$, $K_b = K_{b+} + K_{b-}$, $K_e = K_{e+} + K_{e-}$ are the stiffness terms due to shoulder, biarticular, and elbow muscles, respectively. This equation illustrates that without multiarticular reflexes, the stiffness matrix and thus the couplings between the joints are symmetric. However, multiarticular reflexes could, in principle, allow the CNS to produce asymmetric stiffness or impedance. Monoarticular muscles and reflexes influence the diagonal elements of the stiffness matrix and enable modulation of stiffness at the endpoint along the axes determined by the transformation represented by the Jacobian $\mathbf{J}$. The biarticular muscles produce cross-terms enabling modification of the stiffness geometry along different axes. We will discuss in chapter 8 how the CNS can use this freedom to modify the impedance geometry. Note that this simple investigation has neglected not only the multiarticular reflexes but also differences in the moment arms, which play an important role in the relative contribution to impedance by different muscles, as well as the physiological range over which each muscle can modulate its stiffness. All of these factors must be considered when determining the range over which the CNS can actually modulate the impedance (Hu, Murray & Perreault, 2012).

Table 5.1 summarizes the transformations from sections 5.3, 5.4 and this section. The arrows indicate in which directions the transformations can always be performed. A necessary condition for the existence of an inverse transformation is that the direct transformation involves regular matrices, as is the case for the 2-joint model of planar movements in most of the workspace where the determinant of these matrices in nonzero. In this case $\mathbf{x}, \mathbf{q}, \boldsymbol{\tau}, \mathbf{F}$ are 2×1 vectors (figure 5.3), $\mathbf{K}, \mathbf{K}_x$ are 2×2 matrices, and the velocities and forces, as well as the stiffness and damping matrices, can be inverted in most of the workspace. On the other hand, with the 2-link 6-muscle model, it is not a priori clear how the six-dimensional muscle force vector $\boldsymbol{\mu}$ should be computed to correspond to a given

Table 5.1
Kinematic transformations between hand, joint, and muscle spaces

Hand		Joint		Muscle	
$\mathbf{x}$	$\longleftarrow$	$\mathbf{q}$	$\longrightarrow$	λ	position
$\dot{\mathbf{x}}$	$\dfrac{\mathbf{J}}{\longleftarrow}$	$\dot{\mathbf{q}}$	$\dfrac{\mathbf{J}_\mu}{\longrightarrow}$	$\dot{\lambda}$	velocity
$\mathbf{F}$	$\dfrac{\mathbf{J}^T}{\longrightarrow}$	τ	$\dfrac{\mathbf{J}_\mu^{\,T}}{\longleftarrow}$	μ	force
$\mathbf{K}_x$	$\dfrac{\mathbf{J}^T\mathbf{K}_x\mathbf{J}+\dfrac{d\mathbf{J}^T}{d\mathbf{q}}\mathbf{F}}{\longrightarrow}$	$\mathbf{K}$	$\dfrac{\mathbf{J}_\mu^{\,T}\mathbf{K}_\mu\mathbf{J}_\mu+\dfrac{d\mathbf{J}_\mu^{\,T}}{d\mathbf{q}}\mu}{\longleftarrow}$	$\mathbf{K}_\mu$	stiffness
$\mathbf{D}_x$	$\dfrac{\mathbf{J}^T\mathbf{D}_x\mathbf{J}}{\longrightarrow}$	$\mathbf{D}$	$\dfrac{\mathbf{J}_\mu^{\,T}\mathbf{D}_\mu\mathbf{J}_\mu}{\longleftarrow}$	$\mathbf{D}_\mu$	damping

two-dimensional joint torque vector τ. There are many possible solutions. For example, one can coactivate a pair of antagonistic muscles to produce the same force without modifying the torque at the common joint (assuming the same moment arm). Also, it is not clear which muscles should be activated to produce a torque at the shoulder or elbow, as biarticular muscles can contribute to the torque at either of the two joints. We will analyze how the CNS may deal with this issue in section 5.9.

5.7 Impedance Geometry

In principle, impedance depends on position, velocity, and muscle activation. In order to examine how impedance varies with movement direction, one can measure the restoring elastic force to small displacements of the hand in different directions (Burdet et al., 2000). To visualize the impedance geometry, one plots ellipsoids corresponding to the force caused by a unit displacement in each direction (Mussa-Ivaldi, Hogan & Bizzi, 1985). For example, a *stiffness ellipsoid* defined by the set

$$\left\{ \mathbf{F} = \mathbf{K}_x \frac{\mathbf{x}}{|\mathbf{x}|}, \forall \mathbf{x} \right\} \tag{5.22}$$

is illustrated in figure 5.6. Note that in general the force produced by a perturbation is in a slightly different direction than that of the perturbation.

Figure 5.7 shows how the inertia, viscosity, and stiffness geometry of the arm depend on the (static) posture and on the applied force. The inertia as well as the viscoelastic

A

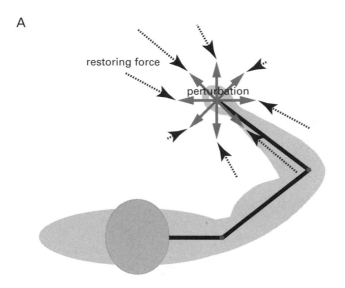

restoring force

perturbation

B
stiffness ellipse

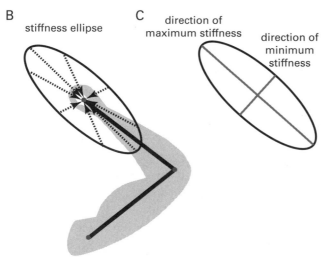

C
direction of
maximum stiffness

direction of
minimum
stiffness

Figure 5.6
Stiffness geometry can be visualized using ellipses corresponding to the force associated with a unit displacement
in different directions. (A) Perturbations are induced in eight directions and the respective restoring force in each
direction is measured. (B) This restoring force can be used to determine the endpoint stiffness of the limb and
represented by an ellipse. (C) This geometric representation enables simple characterization of directional dependence of the magnitude of the elastic response.

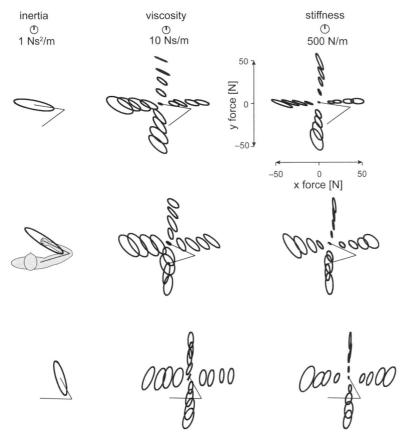

Figure 5.7
Inertia, viscosity and stiffness ellipses of a subject maintaining a constant force with the arm at shoulder height (Perreault, Kirsch & Crago, 2004). Stiffness and damping increase with force but not inertia, which depends mainly on the geometry of body segments.

components of the impedance change when the kinematic configuration of the limb is modified, without requiring any change in muscle activation. We note that the long axis of the inertial ellipse is roughly along the forearm. This seems intuitive, as this direction involves moving both the upper arm and forearm, while the perpendicular direction corresponds to moving only the forearm. The stiffness and damping ellipses are in the same direction, both roughly aligned to the shoulder.

Another way to increase impedance and thus robustness to perturbations is to produce more force. Figure 5.7 shows that stiffness and viscosity increase with the applied force, though inertia, which depends mainly on the body segment mass geometry, does not change with force. Similar to the single-joint impedance described in chapter 4, the

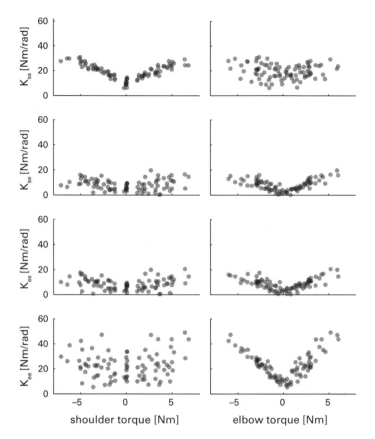

Figure 5.8
Components of joint stiffness $\mathbf{K} = [K_{ss}, K_{se}; K_{es}, K_{ee}]$ of a subject as a function of the torque with the arm in a static posture (Gomi & Osu, 1998). K_{ss} depends mainly on $|\tau_s|$, whereas the other terms depend mainly on $|\tau_e|$.

coefficients of the stiffness matrix grow linearly with the magnitude of applied torque (Gomi & Osu 1998; Perreault, Kirsch & Crago, 2004), as shown in figure 5.8. For example, the mean joint stiffness calculated across the subjects who participated in the study of Gomi and Osu (1998) can be represented as

$$\mathbf{K} = \begin{bmatrix} K_{ss} & K_{se} \\ K_{es} & K_{ee} \end{bmatrix} = \begin{bmatrix} 10.8 + 3.18|\tau_s| & 2.83 + 2.15|\tau_e| \\ 2.51 + 2.34|\tau_e| & 8.67 + 6.18|\tau_e| \end{bmatrix} \frac{\text{Nm}}{\text{rad}} \tag{5.23}$$

Although it is logical that K_{ss} depends on shoulder torque and K_{ee} on elbow torque, we would expect from equation (5.21) that the nondiagonal coefficients depend on both shoulder and elbow torque. The fact that they depend mainly on the elbow torque may indicate that elbow and biarticular muscles are controlled as a unit.

Interestingly, the stiffness matrix of equation (5.23), which includes both intrinsic and reflex components, is nearly symmetric. This symmetry is important, as it corresponds to the signature of springlike conservative forces that can be defined by a potential function. The springlike nature of limb impedance may explain some of the amazing capabilities humans have for stable interaction with the environment. In particular, a system that has springlike properties is guaranteed to interact in a stable manner with passive environments (Colgate & Hogan, 1988).

Although in figure 5.7 both stiffness and damping increase monotonically with the applied force, the increase is less than linear in the case of viscosity. In fact, it was shown by Perreault, Kirsch & Crago (2004) that the coefficients of the viscosity matrix are approximately proportional to the square root of the torque magnitude, such that the damping ratio

$$\zeta = \frac{D_x}{2\sqrt{K_x I_x}} \cong 0.26 \tag{5.24}$$

is kept constant. Human joints are *underdamped*; that is, the limb will oscillate in some situations. This behavior is particularly evident in ballistic limb movements in which a damped oscillation is frequently observed around the final position. Damping can be artificially reduced by means of a negative viscous load to amplify these oscillations (figure 4.12; Milner & Cloutier, 1993).

5.8 Redundancy

Many tasks involve fewer DOF than are available in the musculoskeletal system. In this case, the CNS has to specify how to coordinate the different limbs or muscles, or which trajectory to adopt, in the space of all possible combinations. Inversely, this redundancy enables the CNS to adopt a posture particularly appropriate to a given task. In general, a system is *redundant* when there are more DOF in the device than are required to accomplish the task. We have pointed out that the human arm has at least seven DOF, though only six DOF are required to move in free space (as demonstrated by the fact that most industrial robots have six DOF). This "excess" of DOF relative to task constraints is referred to as *kinematic redundancy*. Also, most movements of the joints can be performed using various muscle combinations, which we refer to as *muscle redundancy*. Finally, in reaching movements, for example, infinitely many paths can be used to reach the target, and the muscles can theoretically be activated with infinitely many different temporal activation profiles to produce different velocity profiles, which we refer to as *trajectory redundancy*.

An example of *kinematic redundancy* is illustrated in figure 5.4A involving the shoulder, elbow, and wrist joints. The kinematics of the linkage are described by the following equation:

$$x' = l_s c_s + l_{se} c_{se} + l_{sew} c_{sew}$$
$$y' = l_s s_s + l_{se} s_{se} + l_{sew} s_{sew}$$

$$s_s = \sin(q_s), \quad s_{se} = \sin(q_s + q_e), \quad s_{sew} = \sin(q_s + q_e + q_w)$$
$$c_s = \cos(q_s), \quad c_{se} = \cos(q_s + q_e), \quad c_{sew} = \cos(q_s + q_e + q_w)$$

$$(5.25)$$

As illustrated in figure 5.4B, there are different possible configurations that allow reaching to a given position $(x', y')^T$; that is, this target can be reached with any orientation of the hand permitted by the range of joint motion. For instance, if the task is to grasp a ping-pong ball, then the arm is redundant, as the ball could be grasped with any orientation of the hand. Another example of redundancy is illustrated by the task of pointing with a laser pointer. In this case, the only constraint is that the laser dot be visible at the desired position on the screen, so the arm has excess DOF because this task can be achieved with different orientations of the hand. On the other hand, if the task requires a more specific orientation, as in lifting a mobile phone from a table, the arm is not redundant. Using the model of equation (5.25) for a specific orientation $\theta = q_s + q_e + q_w$ yields

$$x \equiv x' - l_w \cos(\theta) = l_s c_s + l_{se} c_{se}$$
$$y \equiv y' - l_w \sin(\theta) = l_s s_s + l_{se} s_{se}$$

$$(5.26)$$

which is solved for **q** in equation (5.2).

Muscle redundancy is illustrated in figure 5.1. For example, wrist flexion-extension is controlled by the extensor carpi radialis longus (ECRL), extensor carpi radialis brevis (ECRB), extensor carpi ulnaris (ECU), flexor carpi radialis (FCR), and flexor carpi ulnaris (FCU), which can in principle be activated in an infinite number of combinations to produce the same wrist motion. However, this freedom may be reduced in specific tasks by biomechanical constraints on the muscular system as highlighted in the analysis of finger kinematics (Kutch & Valero-Cuevas, 2012).

In summary, control of the musculoskeletal system is characterized by redundancy (in the joints, in the muscles and in the trajectory) that the CNS must resolve. Conversely, this redundancy may provide advantages to the CNS by allowing it to avoid uncomfortable postures, to produce mechanical impedance adapted to the task (as will be discussed in chapter 8) or to minimize effort (as will be discussed in chapters 6 and 9). Humans generally repeat a task using the same patterns of movement with little variation and in particular coordinate the limbs in a specific way. We will now examine how this can be modeled.

5.9　Redundancy Resolution

It is possible to design robots that are redundant—for example, with more joints than required by the task DOF. In this case, how should the joint angles be specified to follow a trajectory specified in task space, and what flexibility does this offer for different applications? Let us consider the three joint linkage of figure 5.4A and the problem of how to select $(x'(t), y'(t))^T$ to achieve a desired trajectory $\{(q_s(t), q_e(t), q_w(t))^T, t \in [0, T]\}$. We begin by considering the differential kinematics

$$\begin{bmatrix} \dot{x}' \\ \dot{y}' \end{bmatrix} = \begin{bmatrix} J_{11} & J_{12} & J_{13} \\ J_{21} & J_{22} & J_{23} \end{bmatrix} \begin{bmatrix} \dot{q}_s \\ \dot{q}_e \\ \dot{q}_w \end{bmatrix}, \quad J_{ij} \equiv \frac{\partial x_i}{\partial q_j}(\mathbf{q}) \tag{5.27}$$

A method to solve the redundancy is to add a constraint. For example, if we specify the orientation $\theta = q_s + q_e + q_w$, then the differential kinematics become

$$\begin{bmatrix} \dot{x}' \\ \dot{y}' \\ \dot{\theta} \end{bmatrix} = \begin{bmatrix} J_{11} & J_{12} & J_{13} \\ J_{21} & J_{22} & J_{23} \\ 1 & 1 & 1 \end{bmatrix} \begin{bmatrix} \dot{q}_s \\ \dot{q}_e \\ \dot{q}_w \end{bmatrix} \equiv \mathbf{J}\dot{\mathbf{q}}, \tag{5.28}$$

and the joint angles can be computed using

$$\dot{\mathbf{q}} = \mathbf{J}^{-1}(\mathbf{q})\dot{\mathbf{x}} \tag{5.29}$$

if the inverse $\mathbf{J}^{-1}(\mathbf{q})$ is defined (that is, if the condition $\det \mathbf{J}(\mathbf{q}) \neq 0$ is satisfied). A physiological example of redundancy reduction using kinematic constraints is Listing's law, according to which the three eye rotations are combined (Tweed & Vilis 1990).

　　Another approach to reducing redundancy consists of minimizing the speed $\|\dot{\mathbf{q}}\|$ under the constraints of equation (5.3). This minimization is particularly relevant for robots equipped with hydraulic actuators. As the oil flow is roughly proportional to speed, minimizing speed leads to moving less oil and saving energy. The Lagrange undetermined multipliers method gives the solution to this constrained minimization problem as

$$\mathbf{q}^* = \mathbf{J}^\dagger \dot{\mathbf{x}}, \quad \mathbf{J}^\dagger = \mathbf{J}^T (\mathbf{J}\mathbf{J}^T)^{-1}, \quad \det(\mathbf{J}\mathbf{J}^T) \neq 0 \tag{5.30}$$

It can be easily verified that this solution satisfies equation (5.3). The *pseudo-inverse* $\mathbf{J}^\dagger$ is a *right inverse* of $\mathbf{J}$, that is,

$$\mathbf{J}\mathbf{J}^\dagger = \mathbf{J}\mathbf{J}^T (\mathbf{J}\mathbf{J}^T)^{-1} = 1 \tag{5.31}$$

　　In the redundant system of the human body, the limbs can also be coordinated to perform an action while minimizing the joint speed. We will illustrate how this method works with the simple example of a one-joint device actuated by two rotary actuators placed in series. The joint angular velocity is

$$\dot{\theta} = \dot{q}_1 + \dot{q}_2 = \begin{bmatrix} 1 & 1 \end{bmatrix} \begin{bmatrix} \dot{q}_1 \\ \dot{q}_2 \end{bmatrix} \equiv \mathbf{J}\dot{\mathbf{q}} \tag{5.32}$$

The pseudo-inverse is computed as follows:

$$\mathbf{J}\mathbf{J}^T = \begin{bmatrix} 1 & 1 \end{bmatrix} \begin{bmatrix} 1 \\ 1 \end{bmatrix} = 2, \text{ thus}$$

$$\mathbf{J}^\dagger = \mathbf{J}^T \left(\mathbf{J}\mathbf{J}^T \right)^{-1} = \begin{bmatrix} 1 \\ 1 \end{bmatrix} \frac{1}{2} \tag{5.33}$$

and the joint angular velocity vector which minimizes the quadratic joint velocity $\dot{q}_1^2 + \dot{q}_2^2$ is

$$\begin{bmatrix} \dot{q}_1^* \\ \dot{q}_2^* \end{bmatrix} = J^\dagger \dot{\theta} = \begin{bmatrix} \dot{\theta}/2 \\ \dot{\theta}/2 \end{bmatrix} \tag{5.34}$$

Note that the pseudo-inverse of equation (5.30) is in general not integrable; that is, a closed movement may not end in the same configuration from which it started. Mussa-Ivaldi and Hogan (1991) showed how the pseudo-inverse can be modified to avoid this problem and yield an integrable weighted pseudo-inverse.

5.10 Optimization with Additional Constraints

Using redundancy, it is possible to modify a manipulator's configuration without changing the end effector position or orientation. This procedure involves "internal motions" restricted to the manipulator's *nullspace*:

$$N \equiv \{ \dot{\mathbf{q}}, \mathbf{J}\dot{\mathbf{q}} = \mathbf{0} \} \tag{5.35}$$

One can show that if $\dot{\mathbf{q}}_o$ is a particular solution of $\dot{\mathbf{x}}_o = \mathbf{J}\dot{\mathbf{q}}_o$, then, for all $\Delta\dot{\mathbf{q}} \in N$, $\dot{\mathbf{q}}_o + \Delta\dot{\mathbf{q}}$ is also a solution. Therefore, we can optimize a cost $V(\mathbf{q}, \dot{\mathbf{q}})$ by performing movement within the nullspace. In particular, the cost V can be reduced using the gradient descent

$$\Delta\dot{\mathbf{q}} = -\chi \nabla_n V \mathbf{n} = -\chi (\nabla V \cdot \mathbf{n})\mathbf{n}, \quad \chi > 0, \quad \mathbf{n} \in N \tag{5.36}$$

where

$$\nabla_n V \equiv \lim_{\lambda \to 0} \frac{V(\mathbf{q} + \lambda\mathbf{n}) - V(\mathbf{q})}{\lambda}$$

is the derivative in direction $\mathbf{n}$ and

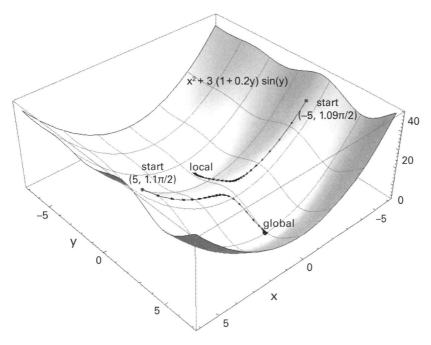

Figure 5.9
Illustration of gradient descent minimization, where the parameter is shifted in direction of the steepest descent of the cost function. Note the convergence to distinct minima depending on the starting conditions.

$$\nabla V \equiv \begin{bmatrix} \dfrac{\partial V}{\partial q_1} \\ \vdots \\ \dfrac{\partial V}{\partial q_N} \end{bmatrix}$$

the gradient, which is projected onto the nullspace by the scalar product with $\mathbf{n}$.

The gradient descent minimization method is illustrated in figure 5.9. As the gradient is in the direction of steepest descent, moving the parameter in this direction tends to a minimum. Although gradient descent is an efficient numerical optimization method, it can get trapped in a local minimum or have slow convergence if the cost function V is relatively flat. Both of these issues can be addressed by adding a stochastic perturbation of the parameter during the search.

We next illustrate how this method can be used to avoid uncomfortable postures. To avoid joint limits $\mathbf{q}^-$ and $\mathbf{q}^+$, one can, for example, tend to remain in the middle $\mathbf{q}^* \equiv (\mathbf{q}^+ + \mathbf{q}^-)/2$ of the range $\{[q_i^-, q_i^+]\}$ in each dimension using the cost

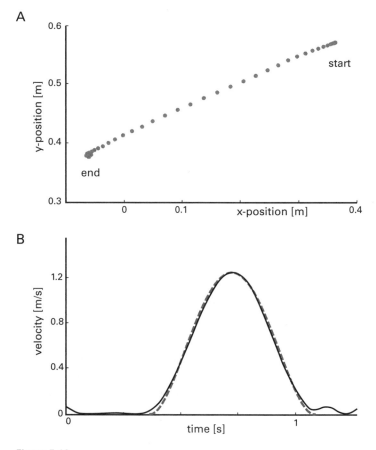

Figure 5.10
Trajectory of a typical planar arm movement, with straight-line path (A) and bell-shaped velocity profile (B). The dashed line in (B) corresponds to a fit with the polynomial of equation (5.38).

$$V(\mathbf{q}) = \|\Delta \mathbf{q}\|^2 = (\Delta \mathbf{q})^T \Delta \mathbf{q}, \quad \Delta \mathbf{q} \equiv \mathbf{q} - \mathbf{q}^* \tag{5.37}$$

Gravity and other tasks constraints can also be considered within this framework by adding a corresponding potential function.

These techniques enable us to simulate arm movements. Despite an infinite number of possible ways of grasping an object, observations suggest that humans consistently employ similar postures and paths. In particular, during reaching the hand trajectory is usually quite straight with a bell-shaped velocity profile v (figure 5.10), which can be modeled by a fourth-order polynomial:

$$v(t) = \left(t_n^2 - 2t_n + 1\right) \frac{30 A t_n^2}{T}, \quad t_n = \frac{t}{T} \tag{5.38}$$

where t is the time, A is the movement amplitude, and T the duration of movement (Flash & Hogan 1985). Furthermore, Cruse, Brüwer, and Dean (1993) proposed that during planar movements, the CNS coordinates the redundant joints so as to produce "comfortable" postures; for example, postures that avoid joint limits. By investigating how humans use the redundant arm in planar movements, they found that the posture adopted by subjects corresponded to minimizing cost functions similar to that of equation (5.37). Furthermore, subjective assessment of posture "comfort" by these subjects yielded measures of discomfort approximately matching those produced by the cost functions (Cruse et al., 1990).

These concepts from robotics and optimization provide a framework for proposing hypotheses about human control of movement with the redundant skeletal system. Coordination between the joints can be implemented using the pseudo-inverse, and comfortable postures require cost functions that correspond to joint limit avoidance as expressed in equation (5.37). Figure 5.11 depicts a simulated extension movement of a three-joint arm with shoulder, elbow, and wrist flexion/extension using these principles. The kinematic parameters are given in table 5.2. The hand moves along a straight path with a velocity profile as in equation (5.38) and a movement duration $T = 1$ s. In figure 5.11A, the motion is well distributed between the joints by the minimization of the squared velocity. However, this minimization does not modify the awkward initial configuration of the wrist which is retained throughout the movement. Figure 5.11B shows that a cost function which penalizes postures close to the joint limits permits repositioning of the wrist away from this "uncomfortable" posture during the movement. The results of Cruse, Brüwer, and Dean (1993) showed that this model captures the main characteristics of the kinematics of redundant planar arm movements.

5.11   Posture Selection to Minimize Noise or Disturbance

In addition to the joint speed, gravity, or uncomfortable postures at the joint limits, redundancy in the musculoskeletal system enables the CNS to minimize the influence of internal or external noise in tasks requiring accuracy. For instance, in order to minimize deviations in hand movements caused by intrinsic tremor, microsurgeons and watchmakers learn to adopt comfortable body postures that attenuate undesired movements. In addition, they support their arms on a solid surface to minimize hand tremor by reducing the number of limb segments subjected to noise.

The theory of uncontrolled manifolds (UCM) (Scholz & Schöner, 1999) uses the nullspace formalism of equation (5.35) to analyze coordination strategies for minimizing the influence of noise. It distinguishes between the interjoint variability in the nullspace of the Jacobian relating joint to endpoint spaces, which does not affect the endpoint position, and orthogonal variability, which affects endpoint accuracy. Experimental evidence

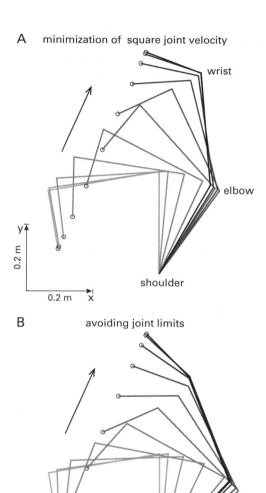

A   minimization of square joint velocity

wrist

elbow

shoulder

B   avoiding joint limits

Figure 5.11
Simulation of redundant arm movement in the horizontal plane at shoulder height. Although minimization of square joint velocity coordinates the joint movements (A), only additional consideration of cost function specifying the joint range enables the arm to avoid the uncomfortable initial wrist configuration close to the joint limit (B).

Table 5.2
Anthropometrical data for arm segments used in the simulations

	Mass [kg]	Length [cm]	Center of mass from proximal joint [m]	Mass moment of inertia [kg m^2]
Upper arm	1.93	0.31	0.165	0.0141
Forearm	1.52	0.34	0.19	0.0188
Hand	0.52	0.08	0.055	0.0003

for UCM is reviewed by Latash, Scholz, and Schöner (2007), including studies on sit-to-stand movements and coordination of finger forces.

A more recent experiment further investigated whether humans select limb configuration to yield an endpoint impedance that avoids instability. While adopting a suitable limb configuration is an energetically efficient way to modulate impedance, do humans spontaneously select the posture most suitable for a given task? Trumbower et al. (2009) designed an experiment to address this question by requiring subjects to stabilize the hand in the presence of a disturbance produced by a robotic interface. The subjects were prompted to explore the space of different postures of the arm and were free to adopt any posture. The disturbance was applied along each of the three Cartesian axes. As shown in figure 5.12, the postures selected by the subjects represented configurations with high stiffness along the direction of the perturbation, thereby attenuating its effect. Subjects typically exploited the passive properties of the arm mechanics to reduce the effort required to counteract the perturbation. For instance, subjects extended the elbow to attenuate the effect of a disturbance in the forward/backward direction; full extension of the elbow represents a singularity in the stiffness (tending toward infinity) along this direction, so little effort is required to oppose a disturbance in this posture. Similarly, subjects flexed the arm to oppose a lateral disturbance. This finding demonstrates that the CNS can select a posture with impedance of the arm adapted to the task.

5.12 Summary

In this chapter, we have covered the following:

• Analysis of complex multijoint multimuscle systems in motor behavior is facilitated by mathematical tools such as infinitesimal kinematics.

• Three coordinate systems are involved in motor control, representing the task, the joint kinematics, and the muscle geometry.

• Mechanical interaction with the environment is controlled through mechanical impedance: the resistance to imposed motion.

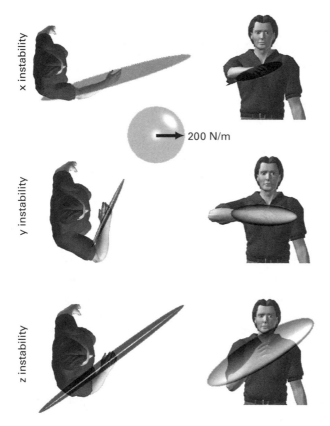

Figure 5.12
Subjects tend to select a posture producing impedance adapted to the task (reproduced with permission from Trumbower et al., 2009). In this experiment, the subject had to maintain the hand at a target position despite a sinusoidal disturbance provided by a robotic interface. This figure shows the posture adopted by one representative subject and the stiffness ellipsoid.

• In a linear model, impedance can be represented by three components—stiffness, viscosity, and inertia—which depend on the body geometry.

Further, we have discussed how to transform position, velocity, force, stiffness, damping, and inertia between coordinate systems. We also covered redundancy in the human body and how the addition of constraints and optimization approaches can be used to resolve control issues arising from redundancy of the skeletal system. Extension of this approach to deal with redundancy in the muscular system is described in the next chapter.

Experimentally, joint stiffness has been shown to increase linearly with the applied torque in static postures while the damping ratio remains constant. Joint stiffness and impedance do not appear to vary significantly with the joint angles, although the endpoint

(Cartesian) impedance of a limb varies dramatically with the joint configuration (due to the position-dependent Jacobian), giving humans the ability to adopt a posture that produces an impedance geometry appropriate to the task they desire to perform.

Although the examples developed in this chapter involved mainly planar arm movements, the techniques are valid for arbitrary movements of the arm in 6DOF environments. In general tasks, care should be taken to parameterize orientation in a suitable way, as is described in standard robotics books, such as Spong, Hutchinson, and Vidyasagar (2006) and Siciliano et al. (2009).

6 Multijoint Dynamics and Motion Control

In the previous chapter, we modeled the kinematics of multijoint systems such as the arm. In this chapter, we will consider the forces and mechanical impedance of multijoint systems during movements performed when interacting with the environment. Let us consider, for example, the movement to reach a cup of tea. When the hand is disturbed by hitting an obstacle along the way, it generally does not deviate much from its undisturbed trajectory and tends to come back to it because of muscle viscoelasticity and neural feedback (Won & Hogan, 1995; Gomi & Kawato, 1997; Burdet et al., 2000). We can analyze the motion control with respect to the undisturbed trajectory by representing the restoring muscle forces as a function of the deviation. For small deviations, we can use a linear approximation which corresponds to the viscoelastic impedance.

The features of human motion control are similar to nonlinear robot control in which movement dynamics are learned and compensated for by the controller and small disturbances cause feedback responses preventing the robot from deviating from the intended trajectory. Therefore, we can use tools from nonlinear control theory to model how the nervous system controls the musculoskeletal system. This approach yields a simple dynamics model that can be used both to simulate movements of animated biomimetic figures (*animats*) or robots and to analyze control of human arm movement and motor adaptation.

6.1 Human Movement Dynamics

In order to perform an action (e.g., move an object on a table), the muscles must accelerate the limb segments as well as the object and compensate for environmental interaction forces (e.g., friction with the table). To perform a limb movement (described in joint coordinates as $\mathbf{q}_u(t), 0 \le t \le T$), the sensorimotor control system must activate selected muscles (represented by the muscle activation vector $\mathbf{u}$) in order to produce a force $\boldsymbol{\mu}\left(\boldsymbol{\lambda}_u, \dot{\boldsymbol{\lambda}}_u, \mathbf{u}\right)$ (where $\boldsymbol{\lambda}_u(\mathbf{q}_u)$ is the muscle length vector corresponding to $\mathbf{q}_u$) for moving the limb segments (with dynamics $\boldsymbol{\tau}_B(\mathbf{q}_u, \dot{\mathbf{q}}_u, \ddot{\mathbf{q}}_u)$) and interacting with the environment. The dynamics equation modeling this interaction in joint space is thus

$$\mathbf{J}_\mu^T\mu\left(\boldsymbol{\lambda}_u,\dot{\boldsymbol{\lambda}}_u,\mathbf{u}_u\right)=\boldsymbol{\tau}_B\left(\mathbf{q}_u,\dot{\mathbf{q}}_u,\ddot{\mathbf{q}}_u\right)-\mathbf{J}^T\mathbf{F}_E \tag{6.1}$$

where $\mathbf{F}_E$ is the force exerted on the hand by the environment. Note that for simplicity the time dependence (t) is omitted in all variables. While consecutive trials of the same task will be similar, they are not identical (as can be seen, for example, in figure 5.5). Therefore $\mathbf{q}_u$ is slightly different for every trial. $\boldsymbol{\tau}_B$, which is the torque required to accelerate the limb inertia, can be modeled using rigid-body dynamics (DeWit, Bastin & Siciliano, 1996) described by

$$\boldsymbol{\tau}_B\left(\mathbf{q},\dot{\mathbf{q}},\ddot{\mathbf{q}}\right)=\mathbf{H}\left(\mathbf{q}\right)\ddot{\mathbf{q}}+\mathbf{C}\left(\mathbf{q},\dot{\mathbf{q}}\right)\dot{\mathbf{q}}+\mathbf{G}\left(\mathbf{q}\right) \tag{6.2}$$

where $\mathbf{H}\left(\mathbf{q}\right)$ is the (position-dependent) mass matrix, $\mathbf{C}\left(\mathbf{q},\dot{\mathbf{q}}\right)\dot{\mathbf{q}}$ represents the Coriolis and centrifugal velocity dependent forces, and $\mathbf{G}\left(\mathbf{q}\right)$ the gravity term. Position and velocity-dependent terms can be added to this equation to account for joint stiffness and damping, respectively.

The dynamics of serial mechanisms such as human limbs can be computed iteratively using algorithms in robotics software packages (e.g., Featherstone, 2008). For example, applying the rigid body dynamics of equation (6.2) to horizontal movements of the two-link model of the arm shown in figure 5.3, we obtain

$$\mathbf{H}\left(\mathbf{q}\right)=\begin{bmatrix}H_{ss} & H_{se}\\ H_{es} & H_{ee}\end{bmatrix},\ \mathbf{C}\left(\mathbf{q},\dot{\mathbf{q}}\right)\dot{\mathbf{q}}=\begin{bmatrix}m_e l_s l_{me}\dot{q}_e\left(2\dot{q}_s+\dot{q}_e\right)s_e\\ m_e l_s l_{me}\dot{q}_e^2 s_e\end{bmatrix}$$

$$H_{ss}=I_s+I_e+m_s l_{ms}^2+m_e\left(l_s^2+l_{me}^2+2l_s l_{me}c_e\right) \tag{6.3}$$

$$H_{se}=I_e+m_e\left(l_{me}^2+l_s l_{me}c_e\right)=H_{es},\ H_{ee}=I_e+m_e l_{me}^2$$

where m_s and m_e are the masses of the upper arm and lower arm, respectively; l_s and l_e the corresponding segment lengths; l_{ms} and l_{me} the distances to the respective centers of mass of the segments; and I_s and I_e the moments of inertia. These dynamics can also be formulated as a linear function of the parameter vector $\mathbf{p}$, which simplifies identification of the dynamics parameters:

$$\boldsymbol{\tau}_B=\boldsymbol{\psi}\left(\mathbf{q},\dot{\mathbf{q}},\ddot{\mathbf{q}}\right)\mathbf{p}=\begin{bmatrix}\psi_{11} & \psi_{12} & \psi_{13}\\ \psi_{21} & \psi_{22} & \psi_{23}\end{bmatrix}\begin{bmatrix}p_1\\ p_2\\ p_3\end{bmatrix},$$

$$\tag{6.4}$$

$$\psi_{11}=\psi_{21}=\ddot{q}_s+\ddot{q}_e,\ \psi_{12}=c_e\ddot{q}_e-s_e\dot{q}_e\left(2\dot{q}_s+\dot{q}_e\right),\ \psi_{22}=c_e\left(\ddot{q}_s+\ddot{q}_e\right)+s_e\dot{q}_s^2,\ \psi_{13}=0,\ \psi_{23}=\ddot{q}_e$$

$$p_1=I_e+m_e l_{me}^2,\ p_2=m_e l_s l_{me},\ p_3=I_s+m_s l_{ms}^2+m_e l_s^2$$

Subject-specific parameters can be identified by using parameter estimation techniques to fit movement trajectories by rigorous system identification (Weiss, Hunter & Kearney,

1988) or by referring to anthropometric tables (e.g., Winter, 2009). The typical parameter values that we have used in our simulations of planar arm movements at shoulder height are given in table 5.2.

6.2   Perturbation Dynamics during Movement

Given a movement with desired trajectory $\mathbf{q}_u$ and dynamics modeled by equation (6.1), we consider the effect of a force perturbation $\Delta\mathbf{F}$ arising during movement that modifies the trajectory to $\mathbf{q}$. The dynamics of this perturbed movement are

$$\boldsymbol{\tau}_\mu(\mathbf{q},\dot{\mathbf{q}},\mathbf{u}) = \boldsymbol{\tau}_B(\mathbf{q},\dot{\mathbf{q}},\ddot{\mathbf{q}}) - \mathbf{J}^T\mathbf{F}_E - \Delta\mathbf{F}_E \tag{6.5}$$

where $\boldsymbol{\tau}_\mu(\mathbf{q},\dot{\mathbf{q}},\mathbf{u}) \equiv \mathbf{J}_\mu^T\boldsymbol{\mu}(\boldsymbol{\lambda},\dot{\boldsymbol{\lambda}},\mathbf{u})$ is the torque vector produced by muscles. In general, muscle viscoelasticity and reflexes make the interaction stable; that is, we can assume that if the perturbation $\Delta\mathbf{F}_E$ is small, $\mathbf{q}$ will remain close to $\mathbf{q}_u$. Therefore, we can use a linear approximation of the restoring force as a function of $\mathbf{e} = \mathbf{q}_u - \mathbf{q}$ in equation (6.5), which gives

$$\boldsymbol{\tau}_\mu(\mathbf{q}_u,\dot{\mathbf{q}}_u,\mathbf{u}) + \mathbf{K}\mathbf{e} + \mathbf{D}\dot{\mathbf{e}} = \boldsymbol{\tau}_\mu(\mathbf{q},\dot{\mathbf{q}},\mathbf{u}) = \boldsymbol{\tau}_T(\mathbf{q}_u,\dot{\mathbf{q}}_u,\ddot{\mathbf{q}}_u) + \Delta\boldsymbol{\tau}_T \tag{6.6}$$

where $\boldsymbol{\tau}_T(\mathbf{q}_u,\dot{\mathbf{q}}_u,\ddot{\mathbf{q}}_u) \equiv \boldsymbol{\tau}_B(\mathbf{q}_u,\dot{\mathbf{q}}_u,\ddot{\mathbf{q}}_u) - \mathbf{J}^T\mathbf{F}_E$ represents the unperturbed task dynamics and $\Delta\boldsymbol{\tau}_T \equiv \boldsymbol{\tau}_B(\mathbf{q},\dot{\mathbf{q}},\ddot{\mathbf{q}}) - \boldsymbol{\tau}_B(\mathbf{q}_u,\dot{\mathbf{q}}_u,\ddot{\mathbf{q}}_u) - \Delta\mathbf{F}_E$ the change in the task dynamics due to the perturbation. The activation dependent stiffness $\mathbf{K}$ and viscosity $\mathbf{D}$ are defined as in equation (5.14) but represented in joint space as

$$\mathbf{K} \equiv \left(-\frac{\partial\tau_i}{\partial q_j} - \sum_k \frac{\partial\tau_i}{\partial u_k}\frac{\partial u_k}{\partial q_j} \right), \quad \mathbf{D} \equiv \left(-\frac{\partial\tau_i}{\partial \dot{q}_j} - \sum_k \frac{\partial\tau_i}{\partial u_k}\frac{\partial u_k}{\partial \dot{q}_j} \right). \tag{6.7}$$

Equation (6.6) describes how the sensorimotor control system implements a movement while interacting with the environment. It is discussed in section 6.4 from the perspective of typical robotics control framework that we describe now.

6.3   Linear and Nonlinear Robot Control

In order to realize a desired trajectory with a robot, it is not sufficient to simply calculate the torque required to move the robot along this trajectory and send torque commands to the motors. Using such a strategy, the robot would not be able to accurately track the desired trajectory because any source of noise, such as friction, unmodeled dynamics, or even simple rounding error in the torque commands, would cause it to deviate from the intended motion. In fact, the robot might even become unstable because there is no mechanism to ensure stability. Therefore, for most robot manipulators the tracking of a desired trajectory is performed using a *linear feedback controller* defined by,

$$\boldsymbol{\tau}_{FB} \equiv \mathbf{K}\boldsymbol{\varepsilon}, \quad \boldsymbol{\varepsilon} = \mathbf{e} + \delta\dot{\mathbf{e}} \quad \mathbf{e} \equiv \mathbf{q}_u - \mathbf{q} \quad \mathbf{K} = \mathbf{K}^T > 0 \tag{6.8}$$

This controller implements simple system dynamics to minimize the tracking error $\boldsymbol{\varepsilon}$ during each time interval. The condition $\mathbf{K} = \mathbf{K}^T > 0$ means that for the error to converge to 0, the matrix $\mathbf{K}$ must be *symmetric* (i.e., $\mathbf{K} = \mathbf{K}^T$) and *positive definite* (i.e., $\mathbf{x}^T\mathbf{K}\mathbf{x} > 0 \ \forall \mathbf{x} \neq 0$). A matrix with positive *gains* on the diagonal and 0 gains elsewhere is generally used for $\mathbf{K}$. The *feedback torque* $\boldsymbol{\tau}_{FB}$ provided by the actuators serves two functions: (1) it drives the plant along the desired trajectory $\mathbf{q}_u$, and (2) it stabilizes it against disturbances.

However, the dynamics of most multilink systems, such as a human or a robot arm, are nonlinear. For example, Coriolis and centrifugal forces grow with the velocity squared. Let us consider, for example, the planar manipulator of figure 6.1, which is composed of six links joined by two metallic plates. Its joints have significant friction that can be modeled by $b_{i1}\dot{q}_i + b_{i2}\text{sign}(\dot{q}_i), i = 1, 2$. It is driven by two DC motors (with moment of inertia I_{mot}) that accelerate the limb, inducing nonlinear effects and dynamic coupling between the links. The dynamics of this device can be modeled (e.g., using the virtual work principle as described in Codourey & Burdet [1997]) as

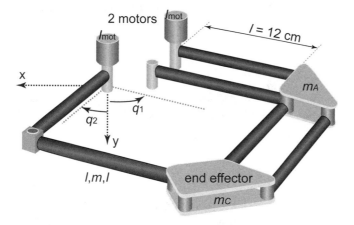

Figure 6.1
Parallel robot with two DOF moving in the horizontal plane. As the motors are not carried out by the structure, this structure can be very light and the robot can move at high speed, yielding nonlinear dynamics that must be compensated for by the controller.

$$\tau = \mathbf{H}(\mathbf{q})\ddot{\mathbf{q}} + \mathbf{C}(\mathbf{q},\dot{\mathbf{q}})\dot{\mathbf{q}} = \psi(\mathbf{q},\dot{\mathbf{q}},\ddot{\mathbf{q}})\mathbf{p}, \; s_{12} \equiv \sin(q_1 + q_2), \; c_{12} \equiv \cos(q_1 + q_2)$$

$$\psi = \begin{bmatrix} s_{12}\dot{q}_2^2 - c_{12}\ddot{q}_2 + \ddot{q}_1 & s_{12}\dot{q}_1^2 - c_{12}\ddot{q}_1 + \ddot{q}_2 \\ \frac{3}{2}\left(s_{12}\dot{q}_2^2 - c_{12}\ddot{q}_2\right) + \frac{11}{4}\ddot{q}_1 & \frac{3}{2}\left(s_{12}\dot{q}_1^2 - c_{12}\ddot{q}_1\right) + \frac{7}{4}\ddot{q}_2 \\ \ddot{q}_1 & \ddot{q}_2 \\ \ddot{q}_1 & 0 \\ \dot{q}_1 & 0 \\ \mathrm{sign}(\dot{q}_1) & 0 \\ 0 & \dot{q}_2 \\ 0 & \mathrm{sign}(\dot{q}_2) \end{bmatrix}^T, \quad \mathbf{p} = \begin{bmatrix} l^2 m_C \\ l^2 m \\ 3I = I_{\mathrm{mot}} \\ l^2 m_A \\ b_{11} \\ b_{12} \\ b_{21} \\ b_{22} \end{bmatrix} \qquad (6.9)$$

which is nonlinear in $(q_1, q_2, \dot{q}_1, \dot{q}_2, \ddot{q}_1, \ddot{q}_2)$, as some coefficients of ψ are trigonometric functions and functions of the squared joint velocity. The dynamics of most manipulators can be described in the same way, that is,

$$\tau = \psi(\mathbf{q}, \dot{\mathbf{q}}, \ddot{\mathbf{q}})\mathbf{p} \qquad (6.10)$$

where the coefficients of ψ are nonlinear functions of $(\mathbf{q}, \dot{\mathbf{q}}, \ddot{\mathbf{q}})$ and $\mathbf{p} = (p_1, \ldots, p_P)^T$ is the *parameter vector* (Codourey & Burdet, 1997). Note that although equation (6.9) is nonlinear in $(\mathbf{q}, \dot{\mathbf{q}}, \ddot{\mathbf{q}})$, it is linear in the parameters, which simplifies their identification.

The linear controller of equation (6.8) cannot compensate perfectly for nonlinear dynamics in all speed regimes because at high speed the torque will increase in a nonlinear fashion. Therefore, a robot with nonlinear dynamics equipped with a linear controller will not be able to accurately follow a dynamically complex trajectory. One solution to this problem consists of adding a nonlinear *feedforward term* $\psi(\mathbf{s}_u)\mathbf{p}$ computed along the desired trajectory $\mathbf{s}_u \equiv (\mathbf{q}_u, \dot{\mathbf{q}}_u, \ddot{\mathbf{q}}_u)$ in order to compensate for the plant dynamics:

$$\tau = \psi(\mathbf{s}_u)\mathbf{p} + \mathbf{K}\varepsilon, \quad \varepsilon = \mathbf{e} + \delta\dot{\mathbf{e}}, \quad \mathbf{e} \equiv \mathbf{q}_u - \mathbf{q}, \quad \mathbf{K} = \mathbf{K}^T > 0, \quad \delta > 0 \qquad (6.11)$$

This *feedforward nonlinear controller*, illustrated in figure 6.2, decouples the two functions previously performed by the linear controller: motion dynamics are provided by the feedforward term, and the *feedback term* $\mathbf{K}\varepsilon$ provides stability and robustness to any error in the feedforward model and to any perturbations arising during movement.

6.4 Feedforward Control Model

To develop a simple model of human arm movements, let us interpret the interaction dynamics of equation (6.6) in the framework of nonlinear control theory. Compensation for the task dynamics along an unperturbed trajectory (i.e., equation [6.1]) is assigned to

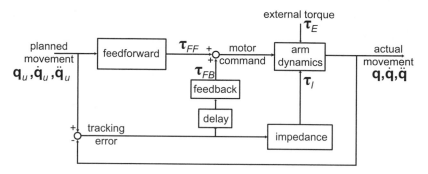

Figure 6.2
Diagram of a feedforward model of motion control in humans.

feedforward control, and the linear terms in equation (6.6) are interpreted as feedback control. In these views, motion control is modeled as

$$\tau = \tau_{FF} + \tau_{FB}$$

$$\tau_{FF} = \psi\left(\mathbf{q}_u, \dot{\mathbf{q}}_u, \ddot{\mathbf{q}}_u\right)\mathbf{p} - \mathbf{J}\left(\mathbf{q}\right)^T \mathbf{F}_E \qquad (6.12)$$

$$\tau_{FB} = \mathbf{K}\left(\left|\tau_{FF}\right|\right)\varepsilon, \quad \varepsilon = \mathbf{e} + \delta\dot{\mathbf{e}}, \quad \mathbf{e} \equiv \mathbf{q}_u - \mathbf{q}$$

This simple *feedforward control model* of human motion is represented in figure 6.2. In equation (6.12), the effect of impedance and neural feedback are lumped, corresponding to the stiffness and viscosity matrices as defined in equation (6.7). The task dynamics depend on the position, velocity, and acceleration whereas the feedforward torque approximating the dynamics depends on the muscle variables, that is, $(\mathbf{q}, \dot{\mathbf{q}}, \mathbf{u})$. An important aspect of this equation is that impedance properties of muscle are implicitly modeled. Similar to equation (5.23), the feedback gain matrix $\mathbf{K}$ is assumed to be a linear function of feedforward torque magnitude. Furthermore, the viscosity matrix is assumed to be approximately proportional to the stiffness matrix. Equation (6.12) extends the linear model of equation (4.11) and yields nonlinear control similar to that employed in robots (equation [6.11]).

Given that the dynamics (equation [6.4]) for a simple two-joint arm model are already complex, it is not obvious how the sensorimotor control system effortlessly produces the torques required to simultaneously move multiple body segments. However, human movement appears to have regular patterns, such that the sensorimotor control system may have to produce only a limited set of movements simplifying the principles required for control. For example, the rigid body dynamics of equation (6.4) scale linearly with time:

$$t \to f(t) \equiv \xi t, \quad \xi > 0 \qquad (6.13)$$

so velocity transforms as

$$\dot{x}(t) \rightarrow \frac{d(x(f(t)))}{dt} = \frac{dx(f(t))}{df}\frac{df(t)}{dt} = \dot{x}(\xi t)\xi$$

and acceleration as

$$\ddot{x}(t) \rightarrow \frac{d(\dot{x}(f(t)))}{dt} = \frac{d\dot{x}(f(t))}{df}\xi\frac{df(t)}{dt} = \ddot{x}(\xi t)\xi^2$$

Therefore, with the exception of the gravity term, which should not vary with time scaling, the dynamics of equation (6.2) scale by ξ^2. This relationship means that to perform a movement faster or slower, the sensorimotor control system could simply scale motion dynamics (Hollerbach & Flash, 1982; Atkeson & Hollerbach, 1985), that is, muscle tension. It has been shown that this basic approach can describe human motion over some speed regimes (Gottlieb et al., 1996) and may thus be useful for the most stereotypical movements.

Similarly, equation (6.12) describes tasks with dynamics that may be expressed as a linear function of a parameter vector $\mathbf{p} = (p_1,\ldots,p_P)^T$:

$$\tau_{FF}(\mathbf{p}) \equiv \mathbf{\psi}\mathbf{p} \tag{6.14}$$

This linear function of activity parameters is analogous to several frameworks proposed to model artificial and biological systems, including the rigid body dynamics model of serial or parallel mechanisms (Codourey & Burdet, 1997), linear artificial neural networks such as described in section 6.9, force fields (Mussa-Ivaldi & Giszter, 1992; Kargo & Giszter, 2008), and time-varying and synchronous muscles synergies (Kamper & Rymer, 2000; D'Avella et al., 2008).

In biological systems, the idea is that this representation corresponds to neural modules. The torque vector of equation (6.14) is replaced by a vector of muscle activations. Each module could consist of a combination of muscles that are activated together in specific ratios, called *muscle synergies*. D'Avella et al. (2008) used a representation similar to equation (6.14) to analyze the EMG signals of eighteen arm muscles during reaching movements to targets in a vertical plane. Six synergies were sufficient to adequately represent movements in all directions at three different speeds. Three *tonic synergies* corresponding to the muscle activation needed to counteract gravity and viscoelastic forces did not vary with speed. Three *phasic synergies* scaled linearly with the movement duration, in a manner similar to equation (6.13), although we would have expected a quadratic relation as muscle activation corresponds roughly to force. D'Avella et al. (2006) were able to represent the pattern of muscle activation during via-point movements using a linear combination of synergies corresponding to the movement from start position to via point and from via point to end position.

6.5 Impedance during Movement

An interesting feature of the model of equation (6.12) is that impedance during motion is assumed to be similar to the impedance during static posture (e.g., equation [5.23] for planar arm movements), which would simplify its parameter identification. However it is not clear whether this assumption is correct, as impedance may depend on muscle dynamics and reflexes. To validate this assumption and illustrate the capabilities of this model to predict mechanical impedance, we simulated the experiments of Gomi and Kawato (1997), who studied free horizontal movements at shoulder height.

In our simulation of this experiment (Tee et al., 2004), the arm was supported horizontally such that the influence of gravity could be neglected. The desired trajectory in Cartesian space uses a bell-shaped velocity profile $v(t)$ as described by equation (5.38), so that the longitudinal movement in y corresponds to $\mathbf{v}(t) = (0, v(t))^T$ and the transverse movement in x to $\mathbf{v}(t) = (v(t), 0)^T$. Joint angles $\mathbf{q} = (q_s, q_e)^T$ were then obtained from the Cartesian position $\mathbf{x} = (x, y)^T$ via the inverse kinematics transformation of equation (5.2). The rigid body dynamics $\boldsymbol{\tau}_B$ were computed along the planned trajectory using equation (6.4). To show the effect of movement on the stiffness geometry, we computed stiffness ellipses during movement and compared them with the corresponding static ellipses at the same hand positions (figure 6.3). Snapshots of the stiffness ellipses were taken at regular temporal intervals of 100 ms.

For both transverse and longitudinal movements, stiffness was higher as compared to postural stiffness, as joint torque was larger because of the muscle force necessary to move the limb segments. In the transverse movement, there was a large generalized increase in stiffness, similar to that observed in measurements during movement. In the longitudinal movement toward the body, the increase in stiffness was slight and mainly in the direction of movement. As shown in the lower panels of figure 6.3, the size, shape, and orientation of the computed stiffness ellipses correspond well to experimental data obtained from similar movements (Gomi & Kawato, 1997). The middle column of the lower panel of figure 6.3 also shows the model prediction for a movement away from the body for which there is no comparison data. Note that the stiffness differs for movements toward and away from the body due to transient effects of the motion dynamics.

6.6 Simulation of Reaching Movements in Novel Dynamics

We also tested how the model controller of equation (6.12) would move the arm in a novel mechanical environment by simulating horizontal reaching movements as in the experiment of Shadmehr and Mussa-Ivaldi (1994). In this experiment, the subjects performed 10 cm reaching movements with a duration of 500 ms, from a central start position to targets in eight different directions, i.e., 0°, 45°, . . . , 315° (figure 6.4A). The subjects

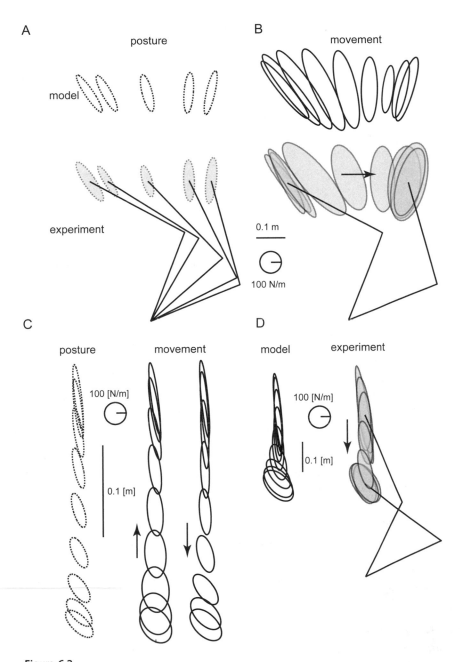

Figure 6.3
Simulation of endpoint stiffness ellipses of the arm in both static postures and during movements. (A) Simulated and experimentally measured stiffness ellipses for static postures of the arm. (B) Simulation and experimental results of transverse arm movements across similar postures to A. (C, D) Simulation and experimental results for longitudinal movements compared with static stiffness. Static and dynamic stiffness predicted by the model compare well with that measured by Gomi and Kawato (1997) where data are available (A, B, D). Reproduced with permission from Tee et al. (2004).

held a handle fixed to a planar robot, which created a *velocity-dependent force field* (VF) during the movement as defined by

$$\begin{bmatrix} F_x \\ F_y \end{bmatrix} = \begin{bmatrix} -10.1 & -11.2 \\ -11.2 & 11.1 \end{bmatrix} \begin{bmatrix} \dot{x} \\ \dot{y} \end{bmatrix} \tag{6.15}$$

where the force is in N and the velocity in m/s.

Simulation of this experiment was performed using the anthropomorphic parameters of table 5.2, stiffness depending linearly on torque as in equation (5.23), and damping proportional to stiffness and decreasing with speed as suggested by isolated muscle studies (figure 3.9):

$$\kappa = \frac{0.42}{1 + \dot{\mathbf{q}}^T \dot{\mathbf{q}}} \tag{6.16}$$

Figure 6.4B shows the effect of the VF on (straight) reaching movements to the eight targets of the Shadmehr and Mussa-Ivaldi study. We see that movements are greatly perturbed by the introduction of the force field compared to the perturbation-free movements produced by the feedforward controller of equation (6.12). The velocity profiles of the perturbed movements have multiple peaks similar to the experimental results. It is interesting that the model of equation (6.12), which is a linear approximation of the interaction dynamics, is capable of capturing the essential features of large trajectory deviations such as those produced by the VF.

6.7 Dynamic Redundancy

If we are planning motion with a device that has more degrees of freedom than required by the task, one important consideration is how to best use this redundancy. For example, how can the sensorimotor control system plan the movements of the body or how can one plan the movements of a humanoid (figure 6.5A,B), in an efficient way? Section 5.9 described how redundancy can be used to minimize the squared joint velocity. When considering redundancy in the context of dynamics, one can instead minimize the kinetic energy:

$$V \equiv \frac{1}{2} \dot{\mathbf{q}}^T \mathbf{H} \dot{\mathbf{q}} \tag{6.17}$$

where $\mathbf{H}(\mathbf{q})$ is the mass matrix (Khatib, 1987). Applying the Lagrange method of undetermined multipliers, it can be shown that the minimization of V under the constraints $\dot{\mathbf{x}} = \mathbf{J}\dot{\mathbf{q}}$ yields

$$\dot{\mathbf{q}}^* = \mathbf{J}_H^{\dagger} \dot{\mathbf{x}}, \quad \mathbf{J}_H^{\dagger} \equiv \mathbf{H}^{-1} \mathbf{J}^T \left(\mathbf{J} \mathbf{H}^{-1} \mathbf{J}^T \right)^{-1} \tag{6.18}$$

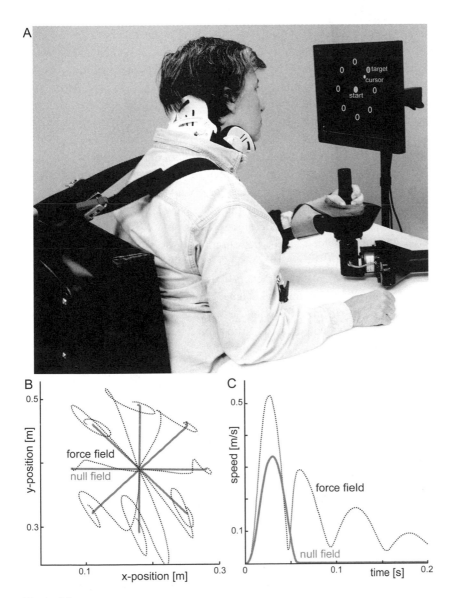

Figure 6.4
Arm movements in a force field. (A) A subject performing reaching movements in eight directions while inter-acting with computer-controlled dynamics produced by the MIT Manus haptic interface (photo by Hermano Igo Krebs). (B) The results of a simulation of a subject performing the same movements as in A in a velocity-dependent force field as defined in equation (6.15) compared to movements without the force field, which is a null field. (C) The velocity profile of the movement in the $-y$ direction in the null field and force field.

where $\mathbf{J}_H^\dagger(\mathbf{q})$ is the *pseudo-inverse* of the Jacobian weighted by the *mass matrix* minimizing the kinetic energy of the rigid-body model. Additional constraints such as gravitation or joint boundaries can be implemented in this scheme using suitable potentials, as was done in equations (5.36) and (5.37). Figure 6.5A shows this method being used to coordinate the movements of a humanoid robot with many degrees of freedom. In this figure, the human experimenter specifies the positions of the robot's hands and the movement is distributed among the joints. Based on this technique, principles can be developed to animate complex figures such as a humanoid in a naturally looking way (Sentis & Khatib, 2005). For example, the comparison of the two postural transitions in figure 6.5B shows how consideration of balance and mobility in a task involving the hands makes the movement of the right animat appear more "natural."

The effect of the minimization of kinetic energy is illustrated in figure 6.5C,D. We simulated a point-to-point movement of the three-joint arm moving horizontally along a straight line with velocity profile of equation (5.38) and duration of one second. Figure 6.5C shows a movement minimizing kinetic energy; figure 6.5D shows, for reference, the same movement of the hand minimizing the square of the joint velocity. The simulation was performed by distributing the movement of the fingertips among the three joints using the weighted pseudo-inverse of equation (6.18) to minimize the kinetic energy and the simple pseudo-inverse of equation (5.30) to minimize the square of the joint velocity. The mass matrix (H_{ij}) was computed (using a software package such as described in Featherstone [2008]) as

$$H_{11} = I_s + m_s l_{ms}^2 + I_e + m_e \left(l_s^2 + l_{me}^2 + 2 l_s l_{me} \cos(q_e) \right) + I_h$$
$$\quad + m_h \left(l_s^2 + l_e^2 + l_{mh}^2 + 2 l_s l_e \cos(q_e) + 2 l_e l_{mh} \cos(q_h) + 2 l_s l_{mh} \cos(q_e + q_h) \right)$$

$$H_{12} = I_e + m_e \left(l_{me}^2 + l_s l_{me} \cos(q_e) \right) + I_h$$
$$\quad + m_h \left(l_e^2 + l_{mh}^2 + l_s l_e \cos(q_e) + 2 l_e l_{mh} \cos(q_h) + l_s l_{mh} \cos(q_e + q_h) \right) = H_{21} \qquad (6.19)$$

$$H_{13} = I_h + m_h \left(l_{mh}^2 + l_e l_{mh} \cos(q_h) \right) + l_s l_{mh} \cos(q_e + q_h) = H_{31}$$

$$H_{22} = I_e + m_e l_{me}^2 + I_h + m_h \left(l_e^2 + l_{mh}^2 + 2 l_e l_{mh} \cos(q_h) \right)$$

$$H_{23} = I_h + m_h \left(l_{mh}^2 + l_e l_{mh} \cos(q_h) \right) = H_{32}, \ H_{33} = I_h + m_h l_{mh}^2$$

with the parameters of table 5.2. The results of figure 6.5C versus figure 6.5D show that when the movement is weighted by the mass matrix, the elbow moves less and the wrist more, as the elbow carries more mass than the wrist.

The technique of equation (6.18) is used to program animats or robots with many degrees of freedom for executing tasks specified in Cartesian space, which has application both in computer graphics simulations and in robotics. Given an endpoint trajectory specified by a task, the trajectories of the joints can be computed using equation (6.18). Figure 6.5A,B illustrates the humanlike quality of motion of a humanoid controlled in this way.

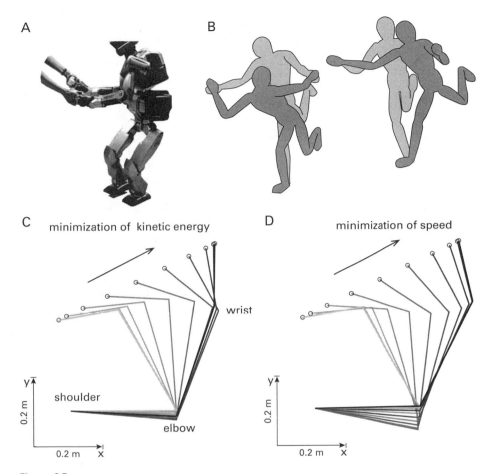

Figure 6.5
Planning the movements of a redundant system such as the human body or a humanoid robot requires specification of principles to coordinate the many joints involved. (A) A complex humanoid's body being taught to follow the experimenter's hands using a pseudo-inverse to coordinate its joints (photo by Abderrahmane Kheddar, CNRS-AIST JRL). (B) The motion of two animats obtained through minimization of kinetic energy as described in section 6.7, with similar foot contact and hand position constraints. The more "natural" posture on the right results from the additional consideration of balance and mobility (based on work by Sentis & Khatib [2005]). The lower panels show the simulation results of reaching with the three-joint arm using minimization of kinetic energy (C) versus square velocity (D).

We have so far considered only kinematic redundancy, that is, how the joints of an animat have to be coordinated. To consider the influence of muscle mechanics and the arrangement of muscles, the square activation

$$V(\mathbf{u}) \equiv \mathbf{u}^2 = \sum_{i=1}^{M} u_i^2 \tag{6.20}$$

can be minimized with the constraints

$$\mathbf{J}^T \mathbf{F} = \boldsymbol{\tau} = \mathbf{J}_{\mu}^T \boldsymbol{\mu} = \mathbf{J}_{\mu}^T \boldsymbol{\psi}(\boldsymbol{\lambda}, \dot{\boldsymbol{\lambda}}, \mathbf{u})$$

As muscle tension $\boldsymbol{\mu} = \boldsymbol{\psi}(\boldsymbol{\lambda}, \dot{\boldsymbol{\lambda}}, \mathbf{u})$ is a nonlinear function of the activation $\mathbf{u}$, the solution to the minimization problem is not as simple as in equation (6.18). However, it can be computed numerically using software packages such as the OpenSim simulator (http://opensimulator.org). The results of Biess, Liebermann, and Flash (2007) demonstrate that this method reproduces the characteristics of human arm movements well. Note that in general the minimization of a cost function (e.g., equation [6.17] or [6.20]) at each point in time does not yield a movement that is optimal over the whole trajectory, that is, one that minimizes

$$\int dt \, \frac{1}{2} \dot{\mathbf{q}}^T \mathbf{H} \dot{\mathbf{q}} \text{ or } \int dt \, \mathbf{u}^2 \tag{6.21}$$

As will be described in chapter 8, humans learn to compensate for dynamics arising from the interaction with the environment, involving compensation for both interaction forces (as expressed in equation [6.1]) and for any instability created by the interaction, such as when using tools. This behavior can be modeled and simulated using the equations

$$\mathbf{J}_{\mu}^T \boldsymbol{\mu} \equiv \boldsymbol{\tau} \equiv \mathbf{J}^T \mathbf{F}$$

$$\frac{d\mathbf{J}_{\mu}^T}{d\mathbf{q}} \boldsymbol{\mu} + \mathbf{J}_{\mu}^T \mathbf{K}_{\mu}(\boldsymbol{\mu}) \mathbf{J}_{\mu} \equiv \mathbf{K}(|\boldsymbol{\tau}|) \equiv \frac{d\mathbf{J}^T}{d\mathbf{q}} \mathbf{F} + \mathbf{J}^T \mathbf{K}_x \mathbf{J} \tag{6.22}$$

where the variables are as defined in chapter 5. We note that specifying the stiffness drastically reduces the system redundancy. Using a dynamics model of the arm, it is then possible to compute the joint torque $\boldsymbol{\tau}$ to perform a movement along a planned trajectory, and using numerical optimization, to determine the muscle activation that satisfies these equations and minimizes some cost function.

6.8   Nonlinear Adaptive Control of Robots

The model of human motor control of equation (6.12) assumes that the feedforward term approximates the motion dynamics. However, these dynamics can change, as during infancy. From daily experience, we also know that humans continually adapt their movements to changing conditions. This *motor adaptation* will be discussed in the next two chapters. Here we present computational methods that are useful for modeling and interpreting the results of experimental studies of motor adaptation. From the perspective of the feedforward control model of equation (6.12), motor adaptation means that the model of the task dynamics must be adapted after each movement.

Let us describe how such adaptation is performed in robotics. Certain tasks performed using robots such as high-speed laser cutting or milling require accurate control throughout the entire movement. When these tasks are performed quickly, it is necessary to compensate for the robot's nonlinear dynamics in order to ensure sufficient precision. In the 1980s, two approaches were developed to learn the robot or task dynamics during motion: *iterative control* (Bien & Xu, 1998) and *nonlinear adaptive control* (Slotine & Li, 1991).

Iterative control corresponds to the typical case of an industrial robot repeating one movement over and over. Using only linear feedback with sufficiently high gains that do not exceed the range of stable control of the robot produces a trajectory that is close to the planned trajectory, which means that the feedback $\tau_{FB}(t)$, $t \in [0,T]$ provides the motion dynamics needed to drive the manipulator along the planned trajectory. The idea of iterative control consists of incorporating a portion of this feedback command iteratively into the feedforward command as the movement is repeated (figure 6.6):

$$\tau_{FF}^{k+1}(t) = \tau_{FF}^{k}(t) + \alpha \, \mathbf{filter}\left(\tau_{FB}^{k}(t)\right), \ 0 < \alpha < 1, \ 0 \le t \le T \tag{6.23}$$

where k is the repetition index. This procedure is repeated over and over until the error becomes minimal. The filter in equation (6.23) prevents high-frequency components from interfering with the learning.

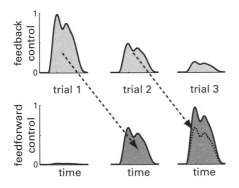

Figure 6.6
Iterative control approach to learning the feedforward command by incorporating a portion of the feedback command after each trial.

Although iterative control is a very simple and efficient method to compensate for the dynamics in repeated movements (Burdet, Codourey & Rey, 1998), an alternative method is required to learn a dynamics model that is valid for arbitrary movements. Iterative control incorporates the feedback command experienced in one movement into the feedforward command for the next movement so that the feedback component becomes negligible. Similarly, nonlinear adaptive control is designed to adapt the feedforward command by updating the parameters of the model in equation (6.10) online in order to minimize the feedback component of equation (6.11). This minimization can be done by gradient descent of the squared feedback motor command

$$V \equiv \frac{1}{2} \boldsymbol{\tau}_{FB}^T \boldsymbol{\tau}_{FB} \tag{6.24}$$

yielding the *adaptation law*

$$\mathbf{p}^{k+1} = \mathbf{p}^k - \alpha \frac{dV}{d\mathbf{p}} = \mathbf{p}^k + \alpha \boldsymbol{\psi}(\mathbf{p})^T \boldsymbol{\tau}_{FB}, \quad \alpha > 0 \tag{6.25}$$

The last equation follows from equation (6.11) with $\boldsymbol{\tau}_{FB} = \mathbf{K}\boldsymbol{\varepsilon}$ and $d\boldsymbol{\tau}/d\mathbf{p} \equiv 0$:

$$-\frac{dV}{d\mathbf{p}} = -\left(\frac{d}{d\mathbf{p}}(\boldsymbol{\tau} - \boldsymbol{\psi}\mathbf{p})^T\right)\boldsymbol{\tau}_{FB} = -\frac{d\boldsymbol{\tau}^T}{d\mathbf{p}}\boldsymbol{\tau}_{FB} + \frac{d(\mathbf{p}^T\boldsymbol{\psi}^T)}{d\mathbf{p}}\boldsymbol{\tau}_{FB} = \boldsymbol{\psi}^T\boldsymbol{\tau}_{FB}$$

This adaptation law, together with the feedforward control law of equation (6.11), provides an efficient way to compensate for the dynamics of tasks involving a stable interaction. Figure 6.7 shows the result of implementing this approach with the robot of figure 6.1. While moving along the trajectory of figure 6.7A, the adaptive controller quickly reduced the error to almost 0 (figure 6.7B). Note that the parameter values initially oscillated and evolved more slowly, requiring about one minute to converge (figure 6.7C). The resulting tracking performance was excellent, as the feedback component was generally negligible (figure 6.7D). This adaptive controller has been implemented on various robots with two, three DOF (Burdet, Codourey & Rey, 1998), and six DOF (Burdet, Honegger & Codourey, 2000), requiring the learning of up to twenty-four parameters. It typically reduces the trajectory tracking error $\int dt|e|$ by more than 90 percent.

6.9 Radial-Basis Function (RBF) Neural Network Model

How can the task dynamics be learned when no physical model is available, such as for a human or a robot arm performing a task during interaction with an unknown environment? One possibility consists of using a neural network model, that is, a mapping that

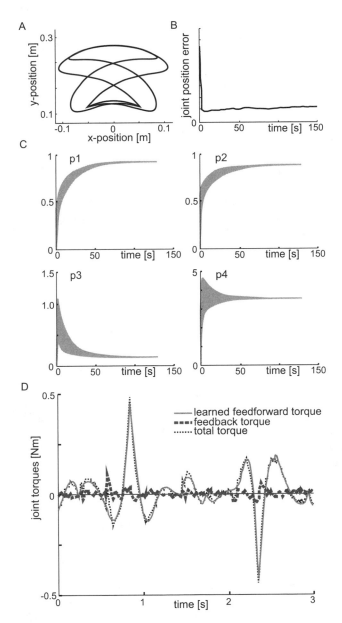

Figure 6.7
Performance of nonlinear adaptive control with the manipulator of figure 6.1 along the trajectory shown in (A). The evolution of integrated tracking error (B) and the first four parameters (C) during learning. (D) The torque after learning. Note that the feedback torque is reduced to almost zero because the nonlinear controller compensates well for the robot's dynamics. Adapted with permission from Burdet and Codourey (1998).

can be adapted during motion. For example, this mapping can be represented as a *radial-basis function (RBF) neural network* described by

$$\boldsymbol{\tau}_{FF} = \mathbf{W}\boldsymbol{\Phi}, \quad \boldsymbol{\Phi} = (\phi_1, \phi_2, \dots, \phi_N)^T, \quad \phi_j(\mathbf{s}) = \exp\left[\frac{\|\mathbf{s} - \mathbf{s}_j\|^2}{2\sigma_j^2}\right], \quad \mathbf{s} = (\mathbf{q}, \dot{\mathbf{q}}) \qquad (6.26)$$

where $\mathbf{W} \equiv (w_{ij})$ are the parameters or *weights* of the neural network that are adapted during learning. $\boldsymbol{\Phi}(\mathbf{s})$ are Gaussian functions representing N (artificial) *neurons* that operate on the state space vector, whose position and velocity components serve as the inputs. Each neuron ϕ_j is characterized by its *center* $\mathbf{s}_j = (\mathbf{q}_j, \dot{\mathbf{q}}_j)$ in the state space and its *activation field* σ_j.

To derive a learning law corresponding to equation (6.25), we first need to express the neural network of equation (6.26) in the format of the linear model of equation (6.14) (Kadiallah, Franklin & Burdet, 2012):

$$\boldsymbol{\psi}\mathbf{p} = \mathbf{W}\boldsymbol{\Phi}, \quad \mathbf{W} = (w_{ij}) \qquad (6.27)$$

Thus

$$\boldsymbol{\psi}\mathbf{p} \equiv \begin{bmatrix} \phi_1 & \phi_2 & \dots & \phi_N & 0 & 0 & \dots & 0 & \dots & 0 & 0 & \dots & 0 \\ 0 & 0 & \dots & 0 & \phi_1 & \phi_2 & \dots & \phi_N & \dots & 0 & 0 & \dots & 0 \\ & & & & & & & & \ddots & & & & \\ 0 & 0 & \dots & 0 & 0 & 0 & \dots & 0 & \dots & \phi_1 & \phi_2 & \dots & \phi_N \end{bmatrix} \begin{bmatrix} w_{11} \\ w_{12} \\ \vdots \\ w_{1N} \\ w_{21} \\ w_{22} \\ \vdots \\ w_{2N} \\ \vdots \\ w_{M1} \\ w_{M2} \\ \vdots \\ w_{MN} \end{bmatrix}$$

Using this identity, it can be shown that the learning law of equation (6.25) yields

$$\mathbf{W}^{k+1} = \mathbf{W}^k + \Delta\mathbf{W}^k, \Delta w_{ij}^k = \alpha v_i \phi_j, \quad \alpha > 0 \qquad (6.28)$$

This technique is well suited to modeling human motor adaptation, as it requires no a priori physical model. The dynamics and convergence properties of this neural network adaptive controller were analyzed by Sanner and Slotine (1992).

6.10   Summary

In this chapter, we have introduced techniques to model and simulate the dynamics of the multijoint arm, its nonlinear adaptive control, and muscle or kinematic redundancy. The concept of a real-time controller is particularly important, as it can be used to elucidate the stable and smooth control of human arm motion. These techniques from robotics will be used in the next chapters to interpret human motor control.

A simple model based on feedforward control was introduced to predict endpoint force and impedance during movement along a known endpoint trajectory. This model postulates that the sensorimotor control system compensates for the task dynamics and that stiffness depends linearly on the joint torque magnitude necessary to move the limb and interact with the environment. The accurate predictions of this model suggest that the impedance of a limb exerting a force on the environment does not depend on whether the force is produced to move the arm or to interact with the environment under static or dynamic conditions. It becomes possible, using measurements of stiffness in static postures for various levels of applied force, to predict the stiffness during arbitrary movements following adaptation to a specified dynamic environment. This model is valid for learned movements of the (possibly redundant) arm or leg in three-dimensional space. Although it is not an accurate physiological model and, in particular, it neglects effects of the cross-bridge dynamics, it can be used as a practical tool to animate figures or robots in a human-like way.

The task dynamics of a human or robot arm can be learned using techniques from nonlinear adaptive control theory and neural networks, which were also introduced in this chapter. It is noteworthy that the adaptation law of equation (6.25) can be applied to iterative control (to learn torque or force along a repeated trajectory, equation [6.23]), to nonlinear adaptive control (to learn a parametric physical model, equation [6.25]), and to an artificial neural network, such as represented by radial basis functions, to learn unstructured dynamics (equation [6.28]) because these techniques are based on the same principle of gradient descent feedback error minimization. The only difference in these three nonlinear adaptive techniques is that the model is a lookup table τ_{FF}^{k} for iterative control, the linear representation of equation (6.10) in the case of nonlinear adaptive control, and the linear network structure of equation (6.26) for a radial basis function neural network. These techniques will be used to model particular aspects of adaptive motor behavior in humans in the next chapters.

In humans, learning is the rule rather than the exception. Unlike animals that walk within hours or minutes of being born, humans take months and years to gradually learn to control movement of their body. There appears to be little innate ability to control movement, so humans must use sensory information to learn how to perform movements and achieve motor behaviors through experience. Learning is required not only to acquire new behaviors but also to improve previously developed behaviors throughout our entire lives. This learning is critical during the early years of development but continues to be important as we grow, age, or experience changing conditions or novel environments such as slippery sidewalks and steep ski slopes.

In order to succeed to reach an object, our sensorimotor system needs to activate the appropriate muscles in a sequence that results in appropriately timed forces being applied to the complex skeletal system. Reaching movements in early childhood consist of a series of small movements guided by visual feedback to gradually bring the arm to the object. During development, these movements gradually increase in speed, become more accurate, and require fewer corrective movements to capture the target object (von Hofsten & Rönnqvist, 1993). In order to produce such movements, the sensorimotor system must acquire a feedforward internal model of the limb and body dynamics. Internal models are neural representations of the relationship between motor commands and their effect on movement of the limbs (Kawato, 1999).

During growth and development, the mechanical properties of our motor system change dramatically. Bones grow and muscle mass increases, changing the dynamics of the limbs. Muscle strength increases such that similar neural activation patterns produce larger forces. In addition to changes in the mechanical structure, there are also changes in signal transmission. Nerve conduction delays initially decrease in the first two years after birth as myelination occurs but then increase in proportion to lengthening of the limbs (Eyre, Miller & Ramesh, 1991). Although these changes taper as we reach adulthood, other changes occur as we begin to age. Conduction delays increase (Norris, Shock & Wagman, 1953) and muscle strength decreases (Lindle et al., 1997) due to muscle atrophy (Jubrias et al., 1997) and changes in muscle mechanics (Brooks & Faulkner, 1994; Larsson, Grimby &

Karlsson, 1979). Whereas most of these changes are gradual, sudden changes frequently occur as well. For example, there is an immediate change in the arm dynamics when we grasp and manipulate objects. Similarly, the manner in which our motor system responds to commands changes as our muscles become fatigued (Enoka & Duchateau, 2008; Enoka & Stuart, 1992). To maintain a desired performance, the controller needs to be "robust" to changes in the dynamics of the arm, such as may occur through adaptation of an internal model of the dynamics. This ever-changing nature of the motor system places a premium on our ability to adapt control appropriately.

Although there are some consistent effects between learning novel dynamics (e.g., using a tennis racquet) and learning novel visuomotor transformations (e.g., using a computer mouse), we will focus almost exclusively on the learning of novel dynamics within this section. The results of these two types of experiments should not be confused, as they result from different manipulations, occur through different learning signals, and are learned independently (Flanagan et al., 1999; Krakauer, Ghilardi & Ghez, 1999). Despite these differences, many of the findings with both types of manipulations are consistent, suggesting that similar neural structures may underlie both types of learning. The understanding of the mechanisms underlying the learning of novel dynamics has been extensively expanded in the last twenty years by using robotic interfaces that the subjects grasp during reaching movements. These robotic devices can apply forces to the subject's hand that change as a function of the state of the limb in the workspace. As such, they can be used to generate virtual objects, simulated walls, or smoothly varying force fields that depend on the subject's kinematics.

7.1 Adaptation to Novel Dynamics

Humans excel in the ability to adapt rapidly to variation in the dynamics of their arm as the hand interacts with the environment. Although there were many earlier studies examining human learning and adaptation through a wide variety of techniques, the idea of internal model formation during adaptation to changes in the dynamics of the limb and environment received its impetus from two landmark studies published in 1994 (Lackner & DiZio, 1994; Shadmehr & Mussa-Ivaldi, 1994). Although they used different techniques, the two studies generated similar insights into the learning of variation in limb dynamics.

Shadmehr and Mussa-Ivaldi (1994) asked subjects to make point-to-point movements in a horizontal plane to a selection of targets while grasping a planar robotic manipulandum. When the robot generates no active forces upon the subjects arm, this is termed a *null force field* (NF). Subjects initially made movements in such a NF in order to become accustomed to the motion and any passive dynamics produced by the robotic manipulandum. Once subjects were able to produce straight reaching movements to each of the targets, a velocity-dependent force field (VF) was introduced in which the force that the robotic manipulandum applied to the hand of the subjects was a linear function of the hand velocity. The force field ($\mathbf{D}_x$) (figure 7.1A), specifying the relation between the forces (F_x, F_y) [N] and the hand movement kinematics (in this case, the Cartesian velocity ($\dot{x}, \dot{y}$) [m/s]) was implemented as

$$\mathbf{F} = \mathbf{D}_x \dot{\mathbf{x}}$$
$$\begin{bmatrix} F_x \\ F_y \end{bmatrix} = \begin{bmatrix} -10.1 & -11.2 \\ -11.2 & 11.1 \end{bmatrix} \begin{bmatrix} \dot{x} \\ \dot{y} \end{bmatrix} \tag{7.1}$$

This force field produces forces on the hand that depend on movement velocity in a manner not normally encountered in daily activities, resisting or assisting forward motion and producing forces perpendicular to the motion depending on the direction of movement (figure 7.1B). The initial movements in this force field exhibited large changes in the trajectory (figure 7.1D), which deviated from normal straight hand paths, similar to the simulation of figure 6.4B. However, as subjects repeated movements, the trajectories gradually became straighter, so after several hundred trials they consistently moved along a roughly straight path to the target similar to the original trajectories with no force field (figure 7.1E).

When the force field was later removed (termed *catch trials*), the subjects' trajectories were again disturbed, but this time in the opposite direction compared to the initial movements in the force field (figure 7.1F). These trajectories are referred to as *aftereffects*. These aftereffects showed that the subjects actually changed the feedforward control of joint torques during the movement to adapt to the force field. That is, when the force field was unexpectedly absent, the limb muscles producing the joint torques required to compensate for the expected force caused the movement to deviate in the direction opposite to the expected force. Specifically, these results demonstrate that the sensorimotor control system did not simply stiffen the joints by co-contraction in order to deal with the force field. If that had been the case, then disappearance of the force field would have resulted in straighter movements. It also indicates that the learned compensation is feedforward in origin, as reverting to the initial dynamics, which originally had no perturbing effect, now disturbs the limb trajectories. This *feedforward controller*, which is gradually adapted in repeated movements, may incorporate an internal model of the task dynamics that computes the feedforward motor command to produce a desired action, based on recalled sensory information about limb and environmental dynamics (corresponding to the model in chapter 6).

The second landmark study, by Lackner and DiZio (1994), investigated adaptation to Coriolis forces during reaching. Subjects were placed in a room capable of rotating about a central vertical axis and asked to produce point-to-point movements between illuminated targets. After performing baseline movements while the room was stationary, the subjects stopped voluntary movement (both head and arm motion) and the room began to rotate, increasing in speed for a period of 2 to 3 minutes until a steady state (60°/s) was reached.

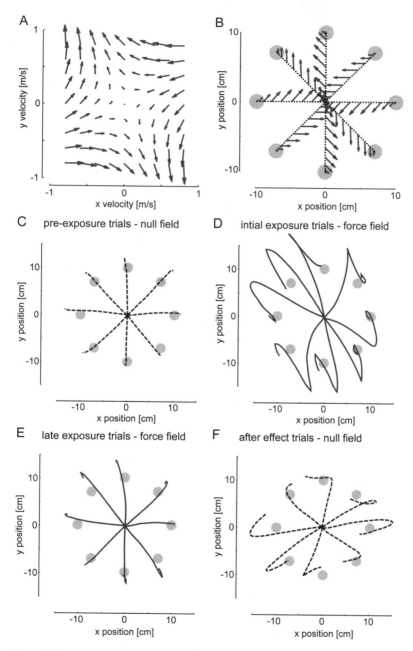

Figure 7.1

Adaptation to novel dynamics. (A) In order to study the learning of novel dynamics, a force field was designed in which the forces experienced in each of the *x*- and *y*- directions depended on the movement kinematics in both of the axes. The forces experienced are plotted as vector arrows indicating the magnitude and direction of the forces as a function of the hand velocity in *x*- and *y*- directions. (B) In the experiments, subjects were asked to make movements in eight equally spaced directions to experience a wide range of the force field. The forces in each of the eight directions (grey arrows) are shown as a function of the eight motions. (C) Subjects made straight movements when the NF was present. (D) Initial movements in the force field were perturbed by the applied forces, causing curved trajectories and corrective movements to the targets. (E) After multiple movements in the force field, subjects were able to make straight movements to the eight targets similar to those made in the NF. (F) When the force field was unexpectedly removed, subjects made curved movements in the opposite direction to the force field (aftereffects) demonstrating that they had learned to compensate for the forces. Adapted with permission from Shadmehr and Mussa-Ivaldi (1994).

After a minute at the constant rotational velocity (allowing for any sensory feedback due to movement in the semicircular canals to disappear), subjects were again asked to make the same reaching movements. The Coriolis forces, arising due to the room rotation, initially disturbed the limb trajectories but again, with practice, subjects were able to adapt to the novel environment and make straight reaching movements to the desired targets. Similarly, once the room rotation was stopped and subjects were asked to make reaching movements, the subjects produced movements with large errors opposite to the ones initially produced at the beginning of learning. These large errors represent aftereffects, indicating that they had learned to change the joint torques rather than stiffening their limbs through muscle co-contraction to reduce error.

The results of these two studies were interpreted as evidence that subjects had developed an *internal model of the new environmental dynamics* and incorporated these dynamics into a previous internal model of the dynamics of arm movements in the original NF. Internal models are neural representations of the relationship between motor commands and their effect on movement of the limbs (Kawato 1999). The internal model is used in computing the feedforward commands to muscles. Together these studies suggest that the sensorimotor system not only forms an internal model of the limb dynamics but can also form a model of the interaction with an external object such as the robotic manipulandum.

7.2   Sensory Signals Responsible for Motor Learning

Human sensory feedback regarding limb movement arises from many different sensory modalities, including force and tactile stimuli, as well as muscle kinematics (changes in length or velocity). Several experimental studies have been carried out to investigate which of these sources of feedback are the main determinants of motor learning. When subjects are adapting to a force field, the inclusion of catch trials (NF trials) induces a transient reversal of learning (Thoroughman & Shadmehr, 2000). The effects are manifested as a transient increase in trajectory error and change in the EMG during force field trials immediately following the catch trial. Scheidt and colleagues (2000) showed that rapid readaptation to the NF requires kinematic error. If a *haptic channel* (i.e., a double-sided virtual wall acting to eliminate kinematic error as an error clamp) was imposed after training in a VF, there was little change in applied force from trial to trial compared to the situation where a NF replaced the VF (figure 7.2). The haptic channel does not resist or assist motion toward the target but limits any movement to either side of the straight line from the initial starting location to the target. Any forces produced by the subject orthogonal to the direction of motion can be measured against the side of the haptic channel using a force transducer. A haptic channel can be produced by creating a one-dimensional spring and damper perpendicular to the direction of motion to the target. For a forward-reaching movement (in the y-axis at $x = 0$), the channel would be calculated as

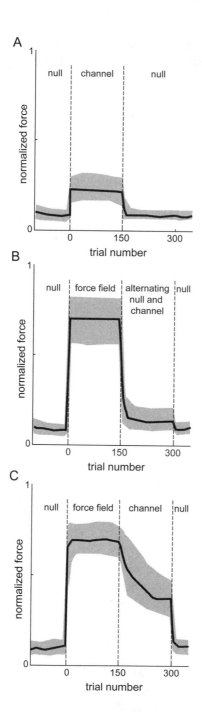

136

Chapter 7

$$\begin{bmatrix} F_x \\ F_y \end{bmatrix} = \begin{bmatrix} K_x x + D_x \dot{x} \\ 0 \end{bmatrix} \tag{7.2}$$

with large stiffness K_x and some damping D_x to avoid vibration (chattering) along the channel.

The results of these studies illustrate that the presence of kinematic error drives the learning process, even if the kinematic error is produced by a NF (Thoroughman & Shadmehr, 2000) or is smaller than 1 cm (Scheidt, Dingwell & Mussa-Ivaldi, 2001). However, when the kinematic error is reduced to almost zero as occurs in the haptic channel, only a small amount of unlearning is observed, as the muscle activity gradually declines trial by trial (Scheidt et al., 2000).

When a force field produces consistent forces on the limb for each movement direction, an internal model of the forces is formed as the kinematic error is systematically reduced. However, if the strength of the force field varies from trial to trial, kinematic error cannot be systematically reduced. In this case, subjects appear to compensate for the approximate mean of the random distribution of the force fields (Scheidt, Dingwell & Mussa-Ivaldi, 2001) driven primarily by the strength of the force field on the most recent trials, that is, by the most recent kinematic errors. In fact, using a model based only on state information from the previous two trials, about 85 percent of the variance during the adaptation process could be accounted for and the size of aftereffects on catch trials could be reproduced. This observation could explain why catch trials lead to a transient increase in error (Thoroughman & Shadmehr, 2000). Together, these studies suggest that the feedforward output of the adaptation process (z) on the subsequent trial $k + 1$ is dependent both on the previous learned output z^k multiplied by a retention factor ϑ and an error based adaptation component. This type of learning can be specified as

$$z^{k+1} = \vartheta z^k + \alpha e^k, \quad \vartheta, \alpha > 0 \tag{7.3}$$

Figure 7.2
Kinematic error is necessary for deadaptation. (A) When subjects perform reaching movements in a NF, the introduction of a mechanical channel causes a small increase in the force at the end of the hand due to forces that would normally cause movement curvature. With enough trials in the mechanical channel, subjects also gradually relax any forces that help constrain the endpoint of the hand along the direction of movement, causing an increase in the endpoint force as the subjects use that interaction to produce any lateral stability. (B) After performing reaching movements in NF trials, subjects adapt to a force field causing a large change in the peak endpoint force. If alternating NF and channel trials are suddenly applied, the peak force produced by the subjects decreases rapidly. (C) However, if after adapting to the force field, channel trials are consistently applied, then subjects only slowly decrease the peak endpoint force. This observation illustrates that without the application of a kinematic error (kinematic error can be introduced through the introduction of NF trials that cause aftereffects), the learned force compensation for the novel dynamics decays slowly.

where the error measure on the last trial e^k is multiplied by a learning rate α. Such a model of learning that attempts to predict the output of the system based on the previous outputs is termed an *autoregressive* model of adaptation.

Error signals that drive motor learning can arise from different modalities such as vision or proprioception. Although learning may progress faster when all modalities are available, it is still possible to learn when sensory information is incomplete. For example, visual feedback of the hand path is not required for adaptation to novel stable (DiZio & Lackner, 2000; Scheidt et al., 2005; Tong, Wolpert & Flanagan, 2002) or unstable dynamics (Franklin, So, et al., 2007). Furthermore, congenitally blind individuals are able to walk and use tools (two examples of adaptation to unstable dynamics) and can adapt to the perturbing effects of a Coriolis force field (DiZio & Lackner, 2000). However, accuracy is compromised when visual feedback is eliminated. When subjects were not provided with any visual information regarding their errors perpendicular to the movement direction, they were able to straighten their movements (adapting to the dynamics) but were unable to adjust movement direction sufficiently to reach the original targets (Scheidt et al., 2005). However, when subjects were provided with visual feedback that eliminated deviations in direction by projecting the hand path onto a straight line joining the start and target locations, they were significantly impaired in correcting for errors in movement direction. These observations suggest that visual information is required for judging the direction of the movement and accurate path planning. Indeed, subjects without proprioception are able to adapt to visuomotor rotations (Bernier et al., 2006), suggesting that vision alone can be used for remapping in movement direction planning. However, subjects without proprioception are unable to learn the correct muscle activation patterns to adapt to self-induced joint-interaction torques during reaching (Ghez, Gordon & Ghilardi, 1995; Gordon, Ghilardi & Ghez, 1995). Thus, visual feedback appears to be required for aspects of learning associated with path planning, whereas proprioceptive feedback is predominately associated with learning and generalization of dynamics.

Changes in the Muscle Activity during Learning

Adaptation to novel dynamics changes the patterns of observed muscle activation. When movements are initially disturbed by force fields, there are large associated changes in muscle activity that are delayed with respect to the onset of movement (Thoroughman & Shadmehr, 1999). Muscle activation increases both in muscles that are stretched and in those that are shortened during the disturbance (figure 7.3). Throughout the force field adaptation period, the field acts in a direction that tends to stretch the extensor muscles (particularly those crossing the shoulder); however, increased muscle activation is seen in all six muscles, both flexors and extensors. This feedback co-contraction is subsequently incorporated into feedforward commands to muscles (Franklin, Osu, et al., 2003) after even a single trial (Milner & Franklin, 2005). Furthermore, it occurs regardless of the size

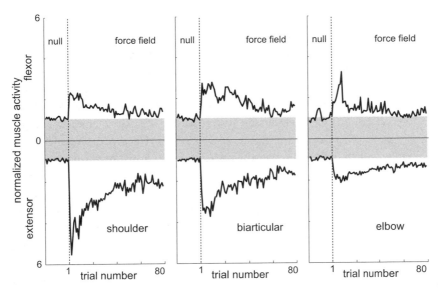

Figure 7.3
Changes in muscle activation during novel force field adaptation. The integrated muscle activity from six arm muscles is shown both before (NF region) and during initial adaptation to a VF. The dotted line indicates the time at which the force field was applied. The gray region illustrates the amount of muscle activity that was required to make the movement in the NF. Muscle activity values have been normalized to the NF levels. Immediately after the onset of the force field, a large increase in coactivation of all muscles is produced. However, this coactivation is gradually reduced as learning progresses, resulting in muscle activity close to the NF levels for the flexor muscles. Based on data from Franklin, Osu, et al. (2003).

of the error (Firouzimehr, 2011). As learning progresses and error is reduced, the feedback muscle activation appears to shift forward in time. The initial feedback to a muscle is gradually reduced and transformed to a feedforward command (Thoroughman & Shadmehr, 1999). At the same time, co-contraction gradually decreases as learning progresses. By the time learning is complete, muscle activity has been modified to effectively compensate for the external force imposed by the force field.

7.3 Generalization in Motor Learning

The studies referred to in the previous sections describe changes to time-varying motor commands learned as the CNS compensates for interaction forces due to the motion of limb segments or interaction forces produced by the environment. One question that arises is whether the CNS needs to explicitly learn the appropriate compensation for every possible movement or whether it can generalize what it has previously learned and apply it to nearby movements or even to movements with different joint configurations (different regions of the reachable workspace).

Evidence of Generalization

What exactly is learned during training in a particular force field in a specific region of the workspace? This question was addressed in the original Shadmehr and Mussa-Ivaldi study (Shadmehr & Mussa-Ivaldi, 1994) in which subjects learned to compensate for a novel force field in one part of the workspace and were tested for aftereffects in a different workspace location. The observed transfer of aftereffects demonstrated that the internal model of the interaction with the environment learned at one specific location could generalize to other regions of the workspace. The next question was whether the internal model generalizes in the task (external) coordinates (Cartesian space) or the limb (internal) coordinates (joint space). Subjects could either be learning the necessary endpoint force to resist the force field or the joint torques that give rise to this endpoint force. The experimenters had subjects learn a force field in one part of the workspace and subsequently produce test movements in another part of the workspace that required a different arm configuration. The force field in this second part of the workspace either represented the same relation between endpoint motion and endpoint force as the previous force field (figure 7.4A), which created a different relation between joint motion and joint torque, or the same relation between joint motion and joint torque, which created a different relation between endpoint motion and endpoint force (figure 7.4B). Specifically, to maintain the same relation in joint space the original force field $\mathbf{D}_x$ (equation [7.1]) was converted to the force field $\mathbf{D}$ by:

$$\mathbf{D}\dot{\mathbf{q}} = \tau = \mathbf{J}^T\mathbf{F} = \mathbf{J}^T\mathbf{D}_x\dot{\mathbf{x}} = \mathbf{J}^T\mathbf{D}_x\mathbf{J}\dot{\mathbf{q}} \quad \forall\dot{\mathbf{q}} \tag{7.4}$$

yielding:

$$\mathbf{D} = \mathbf{J}^T\mathbf{D}_x\mathbf{J} = \begin{bmatrix} 1.66 & 0.64 \\ 0.64 & -1.54 \end{bmatrix} \tag{7.5}$$

where the coefficients of this matrix have units of N·m·s/rad. Whereas maintaining the same relation in endpoint coordinates led to marked perturbation of movement trajectories (figure 7.4C), maintaining the same relation in joint coordinates had almost no perturbing effect (figure 7.4D). This suggests that subjects had formed an internal representation (internal model) of the force field in terms of joint torque and velocity rather than in endpoint force and hand velocity. In that case, the internal representation would be coded in terms of internal coordinates corresponding to the joints or muscles rather than to task space coordinates.

Internal Models Are Coded in State Space

From the previous studies, we know that the sensorimotor control system is able to learn the time-varying motor commands to adapt to novel dynamics and that this mapping from the desired motion to the motor commands—that is, the internal model—is represented in intrinsic coordinates. To investigate whether the internal model implements the structure

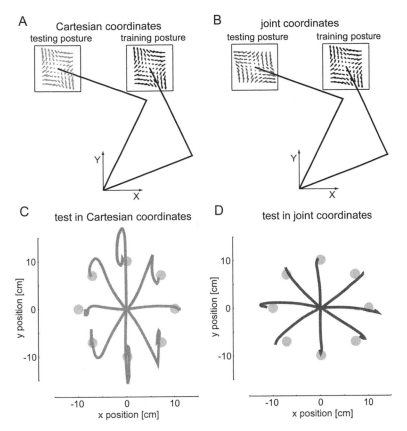

Figure 7.4
Internal models of dynamics generalize in joint coordinates. (A) After subjects have trained in the right hand workspace, they are tested in the left hand workspace. In order to test whether the learning transfers in Cartesian coordinates, the force field in the testing posture is identical in Cartesian space to the force field in the training posture (but different in joint space). (B) In order to test the transfer of learning in joint coordinates, the force field is consistent in intrinsic joint coordinates (joint torque to joint velocity) resulting in a new force field in Cartesian space. The limb postures for the two workspaces are shown with the origin denoting the shoulder joint. (C) Initial reaching movements in the testing posture with the Cartesian coordinate–based force field after previously learning the force field at the training posture. (D) Initial reaching movements in the testing posture with the joint coordinate based force field after previously learning the force field at the training posture. Adapted with permission from Shadmehr and Mussa-Ivaldi (1994).

of the dynamics in terms of the relation between force and state as opposed to memorization of a force profile, Conditt, Gandolfo, and Mussa-Ivaldi (1997) examined whether subjects could generalize the learning of straight line reaching movements to circular movements covering a similar region of the state space. In particular, after subjects had trained by performing point-to-point movements, they attempted to draw circles. To avoid tracing a visually presented shape, the desired circular movement was presented in a different spatial location to the subjects' movement. When the circular movements were performed in a NF, they produced distorted circles, similar to the aftereffects of a different group of subjects who were trained to draw circles in the force field. Furthermore, when the circular movements were performed in the force field after training using only reaching movements, the circles that they drew were essentially identical to those drawn by subjects who had trained using only circular movements. These results provide strong evidence that the CNS does not learn by rote memorization but in contrast forms an internal model (Conditt, Gandolfo & Mussa-Ivaldi, 1997).

Several studies have attempted to examine the structure of the internal model. For example, when we interact with objects in the environment, time-varying forces are applied to our limbs. In principle, the CNS could represent these forces as a temporal sequence or according to their state dependence (e.g., as a function of the limb position or velocity). However, studies have shown that subjects are not able to learn a time-dependent force field (Conditt & Mussa-Ivaldi, 1999; Karniel & Mussa-Ivaldi, 2003). Only when the time-varying forces could be represented in terms of their dependence on state variables such as position and velocity was there generalization of the force field that could be demonstrated by aftereffects. If a time-dependent force field was created that was similar to a force field dependent on state variables, then subjects learned the state variable dependency. The lack of generalization was evident from large errors during movements in directions different from those learned (Conditt & Mussa-Ivaldi, 1999). Therefore, it appears that the neural structures responsible for motor learning rely primarily on state-dependent information.

Additional evidence in support of a state-dependent internal model comes from a study (Goodbody & Wolpert, 1998) in which subjects learned to move in a VF and were subsequently asked to perform movements of half the duration and twice the amplitude as those performed during training. They found that subjects scaled their previously learned model linearly with the velocity in order to extrapolate beyond the velocity used in training, although the extrapolation declined at high speeds. These studies demonstrate that the CNS is able to learn the relation between the limb states and the forces required to generalize throughout state space at least for position- and velocity-dependent states.

Generalization Functions

The question is what the sensorimotor control system learns (what underlying neural functions are being learned by the sensorimotor control system such that generalization occurs).

State space generalization was investigated by Gandolfo et al. (Gandolfo, Mussa-Ivaldi & Bizzi, 1996). Subjects trained by performing reaching movements in two directions separated by 45°. At the end of training, aftereffects were investigated for movements in various directions. In the region between the two directions used in training, and for neighboring directions, strong aftereffects were evident. However, the magnitude of the aftereffects decreased smoothly as the movement directions were displaced farther from the training directions. This observation suggests that the internal model associates state variables and the muscle forces required to produce those states as it is formed. The results of this study show two important features of generalization. The first is that generalization occurs within the state space that has been trained (this can be thought of as interpolation between the previously learned states), and the second is that generalization occurs outside of the previously learned states (extrapolation to new states). However, this extrapolation of generalization decays away from the learned states. Such findings suggest that although the underlying learning functions may be broadly tuned (affecting many neighboring regions), they are still localized within a part of the state space. Thus it appears that the neural structure underlying learning contains local basis functions that combine to produce the continuous adaptation across all of state space.

This question, the determination of the underlying basis functions that produce learning and generalization, was initially investigated by Thoroughman and Shadmehr (2000). By carefully examining the changes in forces and kinematics during adaptation to a VF, they demonstrated that a movement toward a target in one direction affects what has been learned (the internal model) in all other directions. This effect was quantified as the change in error for consecutive movements as a function of the difference in direction between the movements. There was a positive effect (reduction in error) when directions were close, similar to what would have occurred in the movement direction had been the same. The greater the difference between movement directions, the weaker this effect became. For directions that differed by more than 90°, there was a negative effect (increase in error) (figure 7.5A).

In a subsequent study, Donchin, Francis, and Shadmehr (2003) modeled the adaptation using broadly tuned Gaussian-like basis functions and a vector measure of error (Thoroughman & Shadmehr [2000] used a scalar measure of error). When they applied the model to a variety of force fields, they found broadly tuned generalization as in the previous study. However, they no longer found that movements in opposite directions (180°) interfered with the learning. Rather, they found stronger transfer of learning for movements in opposing directions than in perpendicular directions suggesting a bimodal learning function, although the bimodal effect depended on how the error was measured. It was present if the error was measured relative to the trajectory during movements in the NF (figure 7.5B) but disappeared if the error was measured relative to the trajectory at the end of training in the force field (figure 7.5C). Overall, these studies suggest that the neural representation of the dynamics of these novel force fields is similar to broadly tuned

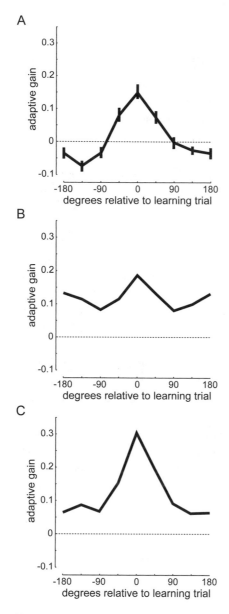

Figure 7.5
Generalization of learning to nearby movements. (A) A movement toward a target in one direction affects what has been learned (the internal model) in all other directions. The amount of learning was quantified as the change in error for consecutive movements as a function of the difference in direction between the movements. The gain of this adaptation function is plotted for different movement directions relative to the previous trial on which the error was produced. Adapted with permission from Thoroughman and Shadmehr (2000). (B) Gain of the adaptive function estimated during force field adaption trials using basis functions with the error measured relative to the trajectory during movements in the NF. Adapted with permission from Donchin, Francis and Shadmehr (2003). (C) Gain of the adaptive function in the curl force field if the error was measured relative to the final trajectory at the end of training in the force field. Adapted with permission from Donchin, Francis, and Shadmehr (2003). All three gain functions were estimated during adaptation to the same curl force field.

Gaussian-like basis functions across the state space. Consequently, compensation for errors occurring in one movement direction modifies the internal model in a way that affects compensation for errors across large regions of the state space.

7.4   Motor Memory

An important feature of motor adaptation is that the acquired internal model persists after training stops; that is, there is no need to continually practice in order to retain the skill. For example, once subjects have learned to compensate for a VF created by a robotic interface, they can go for several days without training, performing many other activities in the interim, and retain the ability to produce relatively straight movements in the same force field (Brashers-Krug, Shadmehr & Bizzi, 1996). This retention of the internal model is termed *motor memory*. It can also be demonstrated by having subjects perform movements in the NF several days after training in the force field. The motor memory is manifested as aftereffects similar to those obtained directly after the period of motor learning. Finally, it can also be found as a faster readaption to the same field after an intermediate set of NF trials (termed *savings*) (figure 7.6A) (Smith, Ghazizadeh & Shadmehr, 2006). One technique used to investigate the processes involved in motor learning is to examine interference and consolidation when switching between different force fields, that is, by examining adaptation to multiple force fields for the same states, rather than generalization to unexperienced states as in the examples described in the previous section. For example, an A-B-A paradigm is very useful for investigating some of these processes. In this paradigm, one force field (field A) is learned first, followed by a completely different force field (field B). Finally, the subjects relearn the original force field (field A). The speed and amount of learning in field A during the second learning session is compared to the first session illustrating a variety of effects that we will discuss in turn.

Interference

Interference can be quantified in terms of the amount by which learning one task interferes with learning another task. In particular, there are two types of interference: *anterograde interference* and *retrograde interference*. Specifically, if we examine learning using the A-B-A paradigm, then retrograde interference is when learning to adapt to force field B interferes with subsequent performance in force field A, to which the subject has previously adapted. Retrograde interference is quantified by comparing the performance in force field A before and after exposure to force field B (figure 7.6B). Anterograde interference is when adaptation to force field A interferes with the subsequent adaptation to force field B. Anterograde interference is quantified by comparing the rate of adaptation to force field B when there is no prior exposure to force field A with the rate of adaptation after having adapted to force field A. In addition to a decrease in the rate of adaptation, anterograde interference also results in greater initial errors.

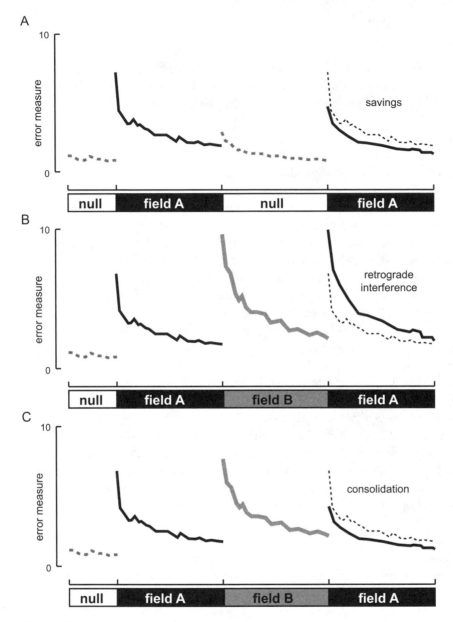

Figure 7.6
Savings, interference, and consolidation in motor learning. (A) Savings demonstrated by learning novel force field A (solid line) followed by NF trials (dotted line). When the force field A is again introduced, the errors are much lower throughout the learning period compared to the original learning curve (dotted line). Savings indicates that some of the original adaptation was retained. (B) Interference is demonstrated by presented force fields in the A-B-A paradigm, that is, force field A trials, followed by force field B trials, followed by force field A trials. In this paradigm, the errors experienced in field B are normally larger than if field B had been learned directly after the NF (anterograde interference). The error measures on the final block of field A trials (solid line) are larger than those experienced on the first block of field A learning trials (dotted line), indicating retrograde interference. (C) Consolidation demonstrates the opposite responses to interference. Here, despite the presentation of the A-B-A paradigm, the errors on the final force field A (solid line) are lower than those of the initial force field A learning (dotted line). Consolidation has been reported when sufficient waiting periods (24 hours) are given between the initial presentation of field A and the subsequent field B presentation.

Interference can be used to investigate the neural processes involved in learning novel dynamics. Given that learning occurs through the development of an internal model of the external environment, interference in attempting to learn two sets of dynamics provides evidence either that the same neural memory or learning structures are being used to learn the internal model of both sets of dynamics or that the learning algorithm is unable to correctly assign the error to the appropriate internal model. For example, the MOSAIC model of learning contains multiple modules of predictor/controller pairs (Wolpert & Kawato, 1998). In such a framework, one of the difficulties is module selection, that is, both determining the relative contribution of each module to any particular contextual situation and associating any errors to the correct module. Thus, interference could arise from using the same predictive/controller pair for two different environments. By exploring which conditions or contexts give rise to an absence of interference, therefore, we can start to understand the way in which the sensorimotor control system partitions learning. For example, it has been shown that VFs and visuomotor rotations exhibit no interference (Flanagan et al., 1999; Krakauer, Ghilardi & Ghez, 1999). Although this finding was originally interpreted as suggesting that dynamics and kinematics are represented in separate neural structures, more recent work has suggested that it is the kinematic variables on which the transformations depend that are more critical in determining the level of relative interference (Tong, Wolpert & Flanagan, 2002). In such an interpretation, a position-dependent field will not interfere with an acceleration-dependent field but two position-dependent fields will interfere with one another.

Consolidation

After subjects have initially adapted to a novel force field, what determines how well the motor memory is retained? The ability to maintain a learned motor memory and avoid disruption through learning of other motor activities is termed *consolidation*. In a series of experiments using an A-B-A paradigm (figure 7.6C), Shadmehr and colleagues (Brashers-Krug, Shadmehr & Bizzi, 1996; Shadmehr & Brashers-Krug, 1997) investigated the consolidation of recent motor learning into long-term memory. By reversing the force field direction (force field B was a mirror image of force field A), they showed that adaptation to force field B immediately after adaptation to force field A destroyed the memory of the force field A when tested on the next day (Brashers-Krug, Shadmehr & Bizzi, 1996). However, if an interval of at least four hours elapsed between adaptation to the two force fields, then the memory of the first force field could be demonstrated twenty-four hours later and for as long as five months. This finding suggests that a certain amount of time is required for the "fragile" short-term motor memory to be consolidated into a more robust long-term memory. One particularly interesting finding relates to the size of the aftereffects at the onset of adaptation to the second force field. If training in the second force field began immediately after adaptation to the first field, then the size of the aftereffects was greater compared to when the second force field was learned 5.5 hours later (Shadmehr & Brashers-Krug, 1997). At the earlier time, formation of the new internal model appears to use the same short-term memory space as the previously learned model causing interference because the CNS is initially using the previously learned model parameters. That there was reduced interference when adaptation to the second force field occurred at the later time suggests that subjects were using short-term memory akin to "tabula rasa"; that is, short-term memory was cleared with passage of time.

However, these findings were not corroborated in subsequent studies using similar experimental designs (Caithness et al., 2004). Regardless of the elapsed time between adapting to one force field (A) and a second reversed force field (B), learning in force field B always interfered with learning in force field A, attenuating this initial memory of the novel dynamics. It has been suggested that these different results may arise from the absence of catch trials in the latter studies (Overduin et al., 2006). Although all of the critical factors necessary for consolidation have not yet been identified, it appears that motor memory is labile like other memory systems (Nader & Hardt, 2009). That is, once a novel skill has been learned, the memory can be retained for long periods of time with little degradation, even without recall.

Spontaneous Recovery and Multiple Learning Processes

By controlling the amount of learning in the second force field with the A-B design, a surprising feature of motor adaptation was demonstrated: the existence of two processes with different time constants (Smith, Ghazizadeh & Shadmehr 2006). In this study, training in force field B was stopped as soon as the aftereffects of adaptation to force field A had disappeared, that is, when the compensatory force exerted by the subjects had essentially reached zero (no force could be detected along a haptic channel) (figure 7.7A). If movements then continued with the haptic channel in place, force against the channel gradually increased in the direction that would have compensated for the first force field (A) rather than force field B (figure 7.7B). This effect (termed spontaneous recovery), initially shown to occur in eye saccadic gain adaptation (Kojima, Iwamoto & Yoshida, 2004), was interpreted as evidence that two learning processes with two different time constants were contributing to the adaptation. The faster process learned and forgot more quickly, whereas the slower process learned and forgot more slowly. This observation was modeled as

$$
\begin{aligned}
z_f^{k+1} &= \vartheta_f z_f^{\,k} + \alpha_f e^k, \quad \vartheta_f, \alpha_f > 0 \\
z_s^{k+1} &= \vartheta_s z_s^{\,k} + \alpha_s e^k, \quad \vartheta_s, \alpha_s > 0 \\
z^{k+1} &= z_f^{k+1} + z_s^{k+1}
\end{aligned}
\tag{7.6}
$$

where x_f is the output of the fast adaptation process and x_s the output of the slow adaptation process. The total compensation (x) is the sum of the two adaptation processes, each of which depends on the adaptation at trial k multiplied by a retention factor (ϑ) and the error during that trial (e^k) multiplied by the learning rate (α). The speed of each of the adaptation modules is specified by the magnitude of the retention factor $(\vartheta_s > \vartheta_f)$ and the learning rate $(\alpha_f > \alpha_s)$.

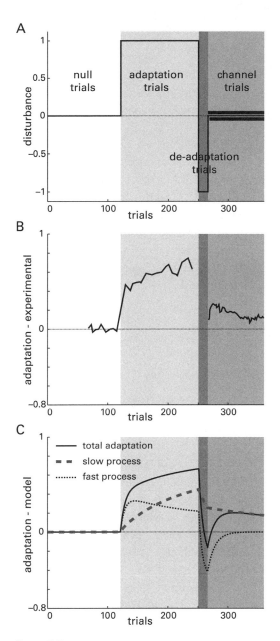

Figure 7.7
Evidence for two adaptive processes with different time constants. (A) Experimental design to demonstrate spontaneous recovery in force field adaptation. After NF trials, subjects learn a force field over multiple movements (adaptation trials). The opposite directed force field is then applied until subjects reduce the total adaptive force to zero (deadaptation trials). Finally, subjects move in a mechanical channel to measure any adaptive force against the channel wall. (B) Experimental data from subjects performing the experiment. The channel trials demonstrate evidence of the increased adaptive force against the channel wall. (C) The predicted adaptation produced by the two state multirate adaption model for the fast process (dotted line), slow process (dashed gray line), and total (solid black line). The spontaneous recovery of the total adaptation can be seen in the channel trials. Adapted with permission from Smith, Ghazizadeh, and Shadmehr (2006).

Initially, the error reduction occurs entirely by means of the fast learning process (figure 7.7C). However, the level of retention attributed to the fast process rapidly reaches a peak and declines thereafter, whereas the slow process, which learns in parallel, gradually leads to greater and greater retention eventually accounting for almost all of the compensation for the force field. If after learning one force field (A) an oppositely directed force field (B) is introduced the fast processes quickly acts to reduce the error. Initially, this occurs by reducing the force needed to compensate for force field A. After a short period of training, that compensating force is effectively reduced to zero. Over this short training interval, the retention attributed to the slow process undergoes relatively little decay. If the error is then clamped (using a haptic channel), the learning generated by the fast process quickly decays, unveiling a significant amount of adaptation to force field A still retained by the slow process. Thus, spontaneous recovery of the adaptation to force field A after a brief interval of interference induced by training in force field B can be ascribed to the slow decay of the original learning by the slow process.

More recently it has been suggested that there may be a single fast process, which is continually engaged to compensate for error, and multiple slow processes, which can be gated by contextual information (Lee & Schweighofer, 2009). Specifically, this has been modeled as

$$z = z_f + \mathbf{z}_s^T \mathbf{c}$$
$$z_f^{k+1} = \vartheta_f z_f^k + \alpha_f e^k \qquad\qquad (7.7)$$
$$\mathbf{z}_s^{k+1} = \vartheta_s \mathbf{z}_s^k + \alpha_s \mathbf{e}^k \mathbf{c}^k$$

where c^k is the contextual cue that switches between all possible slow learning processes that could be associated with the task (figure 7.8). This structure with multiple slow adaptation processes could explain why we can retain independent internal models of sufficiently different tasks without interference (for example, serving the ball in tennis and shooting a free throw in basketball) while more closely related tasks interfere with one another (forehands in both tennis and racquetball). In such closely related tasks, it may be that we cannot identify a contextual cue that allows for switching of the slow process. Some recent evidence has suggested that the fast process may represent error-dependent feedback correction and occur in the cerebellum (Criscimagna-Hemminger, Bastian & Shadmehr 2010) and the slow process may represent formation of predictive feedforward commands, which occurs in the primary motor cortex (Orban de Xivry, Criscimagna-Hemminger & Shadmehr, 2011). This finding may explain why TMS of the primary motor cortex does not interfere with initial force field adaptation but inhibits recall on the subsequent testing sessions (Richardson et al., 2006).

Contextual Effects
Several studies have examined how contextual cues can be used to learn independent models of opposite force fields or visuomotor rotations, which would otherwise interfere.

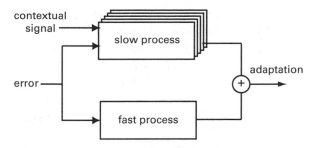

Figure 7.8
Architecture of the single fast and multiple parallel slow learning processes model. Both the learning and the output of the multiple slow processes are gated by the contextual signal so that only one slow learning process is active at any one time. If the contextual signal were modified to partially gate multiple slow processes, then several previously learned slow modules could be adaptively combined to produce responses to novel stimuli. Such a process would require a predictive element, such as paired forward models (Wolpert & Kawato, 1998) for the contextual prediction.

For example, if subjects attempt to simultaneously learn opposing force fields using only the color of the cursor as a cue, no reduction in interference is found (Gandolfo, Mussa-Ivaldi & Bizzi, 1996). That is, the errors at the beginning of the learning and end of the learning are equally large. However, dramatic learning of two opposing force fields has been found when movements in the two force fields require different coordination strategies (Howard, Ingram & Wolpert, 2008, 2010; Nozaki, Kurtzer & Scott, 2006). For example, if training in one force field was a unimanual task and training in the oppositely directed force field was a bimanual task, significant learning in both force fields occurred as measured by aftereffects (Nozaki, Kurtzer & Scott, 2006). Similarly, the ability to retain internal models of oppositely directed force fields is enhanced under conditions in which different neural control mechanisms appear to be involved, such as rhythmic versus discrete movements (Howard, Ingram & Wolpert, 2011).

7.5 Modeling Learning of Stable Dynamics in Humans and Robots

Taken together, the experimental results described in the previous sections suggest several important features of learning dynamics. First, an internal model of the task is formed during learning. Second, this internal model is coded in intrinsic joint coordinates rather than extrinsic Cartesian coordinates. Third, the learning of the internal model is a function of the state and generalizes outside of the learned movements. Finally, the kinematic error is used to adapt this internal model to changes in the dynamics.

As seen in the previous sections, the autoregressive models of adaptation can be used to fit the experimental data. However, as these equations fit only a single error measure on each trial and cannot predict the time-varying changes in the feedforward control, they do not fully capture the learning of an internal model of novel dynamics. For example, they could not be used to simulate or implement entire movements on a robot. What algorithms might capture this mechanism by which the sensorimotor control system adapts to novel dynamics? One possibility is that the feedforward control (internal model) is adapted through the incorporation of the time-varying feedback error signals on a trial-by-trial basis (Albus, 1971; Ito, 1984; Kawato, Furukawa & Suzuki, 1987; Marr, 1969). A similar scheme was introduced in robotics (as described in section 6.8), which might be used as a model of motor adaptation. In this framework, the feedback control signal already provides motion dynamics close to the desired trajectory. The idea of iterative control thus consists of incorporating this feedback signal into the feedforward signal in the next movement (or trial). This procedure is repeated from one trial to the next, until the error, and therefore the feedback, becomes minimal.

Iterative Control in Learning Stable Dynamics

It can be seen, therefore, that iterative learning results in steady decreases in the errors and improvements in the trajectory tracking as learning progresses. A similar change in the patterns can also be found in human adaptation to novel force fields. We can observe the trial-by-trial changes in the trajectories when humans are exposed to a novel force field. Specifically, if subjects are introduced to a VF defined as

$$\begin{bmatrix} F_x \\ F_y \end{bmatrix} = \begin{bmatrix} 13 & -18 \\ 18 & 13 \end{bmatrix} \begin{bmatrix} \dot{x} \\ \dot{y} \end{bmatrix} \tag{7.9}$$

where the force is in N and the velocity in m/s, their trajectories will be disturbed as shown in figure 7.9A (Osu et al., 2003). The trajectories, which initially deviate principally to the left, monotonically converge to the NF trajectories from one trial to the next. This finding suggests that human motor learning corresponds to the iterative learning algorithms of previous section. Indeed, when simulations of iterative learning are performed on the same dynamics, they result in patterns similar to the experimental data, demonstrating the plausibility of this learning algorithm as a mechanism of human motor learning (figure 7.9B) (Burdet et al., 2006). When simulating human movements, the change in the feedback activation signal must be time advanced in order to deal with the delayed neural feedback. The human controller may use kinematic error to update the feedforward model of the task dynamics using iterative learning. Although the motion patterns during simulated learning are similar to the behavior observed in experiments, we have so far only demonstrated the similarity in trajectories for a stable type of mechanical interaction. Further refinement of this algorithm is discussed in chapter 8 on the learning of unstable dynamics.

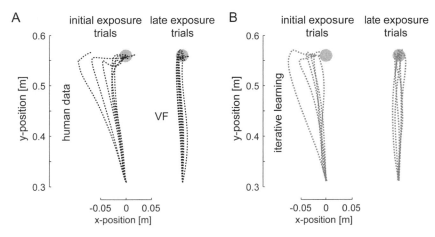

Figure 7.9
Iterative learning of novel dynamics. (A) Reaching movement trajectories for a human subject for initial movements in a novel VF and after learning. Adapted with permission from Osu et al. (2003). (B) Simulated reaching
trajectories in the VF using iterative learning. Iterative control learns the feedforward command by incorporating
a partial amount of the feedback on the previous trial, resulting in a progressively smaller error measure and
therefore feedback command. Adapted with permission from Burdet et al. (2006).

7.6 Summary

To investigate human motor control, one can study the adaptation of movements to novel
dynamics. In order to produce truly novel dynamics, a robotic interface that can generate
and instantly switch artificial dynamics is often used.

The results of these experiments suggest that a feedforward model of the task is formed
during learning using mainly the kinematic error, which appears to be coded in intrinsic
coordinates, and is a function of a state space rather than rote memorization. As such, this
internal model generalizes both inside and outside of the state space in which it has been
learned.

The learning of the internal model occurs through multiple processes: one fast but
general learning process and multiple slow processes gated by contextual information.
This contextual information is essential both for the appropriate internal model selection
for control and for assigning error in order to improve learning. Without such information,
interference between different dynamics occurs, resulting in only the final dynamics being
learned.

Iterative learning algorithms, incorporating feedback into the feedforward model on a
trial-by-trial basis, are able to reproduce many of the experimental results, suggesting that
similar learning mechanisms may underlie motor adaptation.

8 Motor Learning under Unstable and Unpredictable Conditions

Adaptation to novel *stable dynamics* (stable interactions between the human limb and the external environment) requires that the sensorimotor system learn a new mapping between joint torques and joint kinematics. We have discussed in chapter 7 how such tasks are relatively simple to learn, as the forces required can be identified from consecutive movement errors, leading to direct compensation. For example, we may experience difficulty the first time we open a door with high resistance but smoothly perform the action by the second or third attempt. However, many tasks that humans perform—particularly those involving tool use—are inherently unstable (Rancourt & Hogan, 2001). For example, when using a screwdriver, any small deviation in the direction of force applied to the tool can cause the tool to quickly slip in one direction or another. This example highlights that an unstable interaction with the environment, when coupled with the noise inherent in the human neuromuscular system, creates a complex control problem. However, instability is not the only issue that presents difficulties for the learning schemes covered in the previous chapter. Many situations in real life are neither predictable nor repeatable. When walking a dog on a leash, one cannot predict ahead of time the direction in which the dog will pull. From moment to moment, as new stimuli attract the dog's attention, the leash is pulled in different directions. Thus, two critical features of the dynamics that we need to contend with are instability and unpredictability (Franklin & Wolpert, 2011). This chapter explores the manner in which the sensorimotor system deals with these two factors.

Although the sensorimotor control system cannot learn a set of feedforward joint torques to compensate for either of these situations, it can employ other mechanisms to counteract instability and unpredictability. First, the response to disturbances can be reactive, in the form of feedback-driven changes in the motor command. However, feedback pathways are subject to neural delays, ranging from the shortest monosynaptic pathways (conduction delays between 10 and 50 ms, depending on the proximity to the spinal cord) to the much longer reactions times associated with voluntary correction (often longer than 200 ms). In fact, as the corrective action must occur through muscle contraction, the real delay (to the production of muscle force) includes an additional electromechanical delay within the muscle itself that is at least 25 ms for limb muscles (Ito, Murano & Gomi, 2004). Thus, although feedback mechanisms can be used to correct for unpredictable disturbances, if the force produced by a disturbance increases too rapidly, corrective feedback cannot compensate quickly enough to recover from instability.

The second mechanism to counteract instability and unpredictability is based on the principle that muscle activation does not simply change the force but also increases stiffness and viscosity (chapter 3). Therefore, the sensorimotor control system can increase the stiffness and viscosity of joints by muscle co-contraction. This change in the mechanical impedance of the musculoskeletal system is generally stabilizing in nature but requires energy to maintain coactivation. In order to ensure stability but limit energy expenditure, the sensorimotor control system must learn to tune the mechanical impedance of the limbs to environmental instability. Increased limb impedance resists, without delay, both the perturbing effects of any unpredictable disturbance applied by the environment and variability caused by internal or external noise.

8.1 Motor Noise and Variability

Control signals and actuator responses in the human neuromuscular system are contaminated by noise, limiting the accuracy of both our perception and action (Faisal, Selen & Wolpert, 2008). Noise is present at all stages of sensorimotor control, from sensory processing to motion planning and throughout active movement generation. The noise in sensory processing gives rise to errors in estimates of the state of the body (internal state estimates) and errors in estimates of the state of the world around us (external state estimates). This variability directly influences motor output, as predictive control (internal models) is learned as a function of the internal state of the body. Therefore, noise in the internal state estimates directly affects feedforward motor commands, creating differences between the desired motor output based on the current state of the body and the actual motor output based on the sensory system's estimated state of the body.

Noise at the movement planning stage has similar effects to noise at the execution stage, leading to variability in the motion and final position (Gordon et al., 1994; Vindras & Viviani, 1998). Variability in cortical neural firing (in the dorsal premotor area and primary motor cortex) is strongly correlated with future kinematic variability in the movement. In extensively trained reaching movements, variability in peak velocity exhibited a strong positive correlation with variability in the neural firing (Churchland, Afshar & Shenoy, 2006). The results demonstrate that noise in the motor control process is already contributing to movement variability during the earliest stages.

Variability in motor output also results from noise in the motor commands and actuator output (van Beers, Haggard & Wolpert, 2004). The noise in motor commands increases with the size of the motor command (Jones, Hamilton & Wolpert, 2002; Slifkin & Newell, 1999) and is therefore termed signal-dependent noise. As the force increases, the standard deviation of the noise increases linearly with the level of applied force (figure 8.1A). There is evidence that the main contributing factor to the signal-dependent nature of this variability may be the size principle of motor unit recruitment (Jones, Hamilton & Wolpert, 2002). Sensorimotor noise, combined with the dependence of muscle force on activation length change history, and on the current state, creates a situation in which it is virtually impossible for the sensorimotor system to accurately predict motor output.

Thus noise can produce significant variability between consecutive trials with the same goal. This feature of biological motion is illustrated in free reaching movements (figure 8.1B). All trials have the goal of reaching the same target but, despite extensive practice, different trajectories are observed on every trial. Thus, even if one desires to perform the same movement, it will not be possible to do so with the same trajectory.

Interaction between Variability and Instability

Tasks such as opening a door involve a stable interaction with the environment. Thus, similar motor commands result in similar movements, and small perturbations and noise have little overall effect on the movement (Burdet et al., 2006). In contrast, unstable interactions are affected by different initial conditions, neuromotor noise, and small external perturbations, any of which can lead to unpredictable, inconsistent, and unsuccessful performance. These small variations are enhanced by the instability, resulting in large disturbances—large variations in the trajectories. One of the characteristics of instability, therefore, is variance that increases with time.

Take the example of sculpting, in which material irregularities can displace the chisel to the left or to the right of the intended path. The mechanical characteristics of the chisel itself give rise to the instability. Any task involving a tool with a small tip in which forces must be controlled along the length of the tool will be unstable. That is, a small deviation in the direction of the force, or a small movement of the tip, can result in the force being directed outside of the base of support (not along the center of the tool), producing a rotational torque and causing the tool to skip from the desired path. This unpredictable outcome makes unstable tasks difficult to perform and to learn. Consequently, a sculptor requires extensive practice to acquire the skill necessary to compensate for the unstable interaction with the environment.

The contrast between stable and unstable interactions can be seen in the trial-to-trial variability of reaching movements when dynamics are unexpectedly changed (figure 8.2). Noise due to both planning and motor execution results in natural variability of the trajectories. That is, the movement trajectory varies slightly from trial to trial, owing to internal noise (figure 8.2D). If we examine initial movements (prior to learning) in a stable VF (figure 8.2A), we can see similar variation in the movements (figure 8.2C). That is, even before the subjects have learned to compensate for the force field, the interaction between the force field and the human arm exhibits stability; it does not cause enhancement of the natural variations of human movements. Even though the movements are all consistently perturbed by the novel dynamics, causing deviations to the left, every

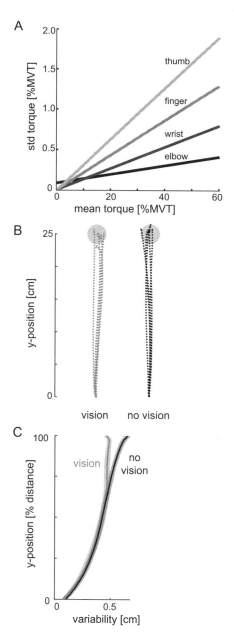

Figure 8.1

Motor noise scales with activation causing variability in the trajectories of reaching movements. (A) The standard deviation of the joint torque scales with the level of joint torque. The size of the variance depends on the size of the muscles involved: higher variance for smaller muscles of the thumb and fingers. Figure based on data from Hamilton, Jones, and Wolpert (2004). (B) Reaching movements with and without vision exhibit variability in their trajectories even though the goal remains constant. (C) The variance of the reaching movement increases throughout the movement, with a small reduction towards the end when visual feedback is present. This finding suggests that although the variability increases with increased motor commands, visual feedback acts towards the end of the movements to limit the variance. Adapted with permission from Franklin, So, et al. (2007).

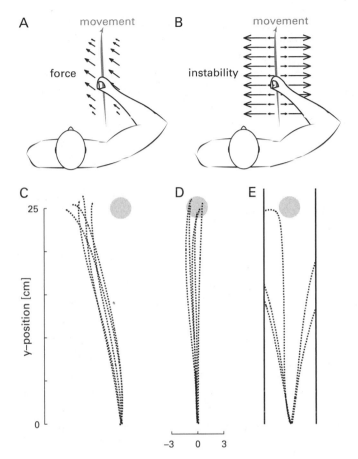

Figure 8.2
Differences between stable and unstable interactions. (A) In the stable VF, the force exerted on the hand is consistent across trials despite small changes in the trajectory. (B) The interaction with the unstable DF is unpredictable due to the presence of motor noise. Although perfectly straight movements will experience no forces, small deviations to either side are amplified by the force field. (C) Movements in the VF (before learning) when applied on random trials among NF trials (termed "before effects"). Although the movements are disturbed, the variability of the movements is similar to those in the NF. (D) Movements in the NF. (E) Movements in the DF when applied on random trials among NF trials (before effects). The field amplifies the natural variability of the movements, causing the trajectories to cross the boundary of the safety zone on either side. Adapted with permission from Osu et al. (2003) and Tee et al. (2010).

trajectory is roughly perturbed with the same profile despite variations caused by the neuromuscular noise. That is, the trajectories on successive trials remain similar despite motor noise. Thus, the same motor command leads to similar movements, making it relatively straightforward to correct for the interaction. This stability can be further demonstrated by applying force pulses on random trials during these movements (Franklin, Burdet, et al., 2003). Despite the application of disturbing force pulses, the trajectories do not diverge further—in fact, the trajectories are virtually indistinguishable several hundred milliseconds after the onset of the force pulse, indicating that the inherent properties of the neuromuscular system could compensate for these pulses within this stable interaction.

Reaching movements can be also performed in an unstable position-dependent divergent force field (DF) such as that defined by

$$\begin{bmatrix} F_x \\ F_y \end{bmatrix} = \begin{bmatrix} K_x x \\ 0 \end{bmatrix} \tag{8.1}$$

where the endpoint forces applied to the subject's hand (F_x, F_y) are determined by the field constant ($K_x > 0$) and the displacement of the subject's hand from a straight-line movement to the target (0, y). The DF produces a negative elastic force perpendicular to the target direction (figure 8.2B), which pushes the hand away from the straight-line trajectory, sometimes left and sometimes right, depending on the initial deviation, causing trajectories to diverge (figure 8.2E). If subjects had been able to perform a perfectly straight motion to the target no force would have been applied to the hand. However, any variations in the hand trajectory were amplified by the instability generated by the force field, resulting in trajectories that quickly crossed the boundaries of a safety zone whereupon the force field was switched off.

To define stability of human movements, we note that the deviation produced by unpredictable disturbances during movement will grow as time passes when the system is unstable but will remain bounded if the system is stable. Given that the naturally occurring noise in the sensorimotor system can be considered as an unpredictable disturbance to movement, the variability of trajectories known to be stable, such as when no force field is present, can serve as a measure of the stability bound. Therefore, we can consider an interaction to be stable if the trajectory variability over consecutive trials is similar to that in NF movements (Burdet et al., 2006).

8.2 Impedance Control for Unstable and Unpredictable Dynamics

Both instability in the environment and unpredictable perturbations can result in trajectory variation. Neville Hogan proposed that the motor control system could adapt the mechanical impedance of our limbs to control the interaction with the environment (Hogan, 1984,

1985). Specifically, he outlined two mechanisms by which this could occur: co-contraction and posture modulation (see chapter 5). Co-contraction of antagonist muscles causes an increase in both the stiffness and viscosity of the joints and therefore also of the endpoint of the limb. Modifying the posture (configuration) of the limb changes both the effective mass of the limb and the postural contributions to endpoint stiffness and viscosity. Through these mechanisms, the sensorimotor control system could tune the impedance of the limb to the instability or unpredictability of the external environment.

In order to test the co-contraction mechanism, Mussa-Ivaldi, Hogan, and Bizzi (1985) applied sinusoidal force perturbations in specific directions to the endpoint of the arm while asking subjects to maintain a fixed posture. The endpoint stiffness of the limb was measured by applying small position-controlled displacements and measuring the restoring force. Subjects modified the stiffness of the limb in an isotropic manner, that is, equally in all directions. This observation suggested that subjects could globally increase muscle co-contraction but could not selectively tune co-contraction to the characteristics of the environment.

However, selective control over the endpoint stiffness of the limb independent of limb posture was later demonstrated for reaching movements in an unstable environment (Burdet et al., 2001). Initial movements in a DF produced trajectories that deviated markedly to one side or the other of a straight line (figure 8.3A). However, with practice subjects learned to make straight reaching movements to the target, with variability similar to that under the NF condition (figure 8.3B). After the period of adaptation, the endpoint stiffness

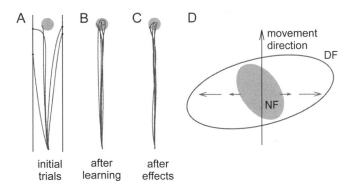

Figure 8.3
Adaptation to unstable dynamics occurs through a selective increase in stiffness in the direction of the instability. (A) The variability of reaching movements is amplified by the instability of the DF, causing initial movements to cross the boundaries of the safety zone (black lines). (B) After learning, subjects are able to make relatively straight reaching movements to the target. (C) After learning the DF, movements performed randomly in the NF (when subjects expect the DF) test for after effects. These after effects are characterized by reduced variability and increased straightness relative to usual movements in the NF. (D) After adaptation, the endpoint stiffness of the arm during the reaching movement is measured using position-controlled perturbations. Relative to the stiffness in the NF (shaded gray ellipse), the DF stiffness (solid line) was increased primarily in the direction of the instability (arrows). Adapted with permission from Burdet et al. (2001).

of the arm, measured using small position-controlled displacements at the midpoint of the reaching movement, was compared to that previously measured in the NF. The endpoint stiffness had increased primarily in the direction of the instability, with little or no increase in stiffness in the orthogonal direction (figure 8.3D), thereby stabilizing the interaction with the environment.

Difference from Stable Adaptation

Although a selective increase in stiffness was demonstrated after adaptation to the DF, a change in endpoint stiffness is also found after adaptation to a stable VF (figure 8.4B) (Franklin, Burdet et al., 2003). The different mechanisms underlying these two types of adaptation were examined by modeling the changes in endpoint stiffness as a function of the change in joint torque (or endpoint force). Although adaptation to the VF requires a distinct change in the endpoint force and therefore net joint torque, adaptation to the DF requires no change in the endpoint force or net joint torque. The change in the endpoint stiffness after adaptation to the VF could be accurately predicted by the change in net joint torque (figure 8.4B). On the other hand, the change in endpoint stiffness predicted by the change in net joint torque after adaptation to the DF was similar to the NF stiffness (figure 8.4A). Thus, the stiffness after VF adaptation was a consequence of the required changes in net joint torque to compensate for the force field. On the other hand, the stiffness change after DF adaptation was not related to changes in net joint torque but due to direct control of the endpoint stiffness through muscle coactivation. This finding was further supported by changes in muscle activity after adaptation (Franklin, Osu, et al., 2003) where DF adaptation exhibited high levels of antagonistic muscle coactivation and VF adaptation exhibited little coactivation.

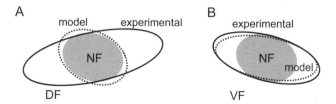

Figure 8.4
Difference in the stiffness after adaptation to stable and unstable environments. (A) The stiffness geometry in the DF predicted by modeling joint stiffness as a linear function of joint torque (dotted ellipse) with both the stiffness measured in the NF (shaded gray) and experimental results in the DF (solid line) superimposed. Stiffness was computed for individual trials and averaged, using the force data for each subject. The ellipses represent averages across six subjects. (B) The modeled (dotted ellipse) and experimental (solid line) stiffness ellipses after adaptation to the VF, along with the NF stiffness (shaded gray). Adapted with permission from Franklin, Burdet, et al. (2003).

Scalability

Endpoint stiffness might be expected to scale with the level of environmental instability. As instability increases or decreases, the sensorimotor control system should adapt the level of endpoint stiffness to ensure stability. Furthermore, if it is efficient, then it should also minimize the metabolic cost. This prediction was tested by examining adaptation to four levels of DF instability and measuring the endpoint stiffness of the limb (Franklin et al., 2004). The endpoint stiffness ellipse scaled with the level of instability (figure 8.5A). The directional selectivity of the stiffness was demonstrated by comparing the components of the endpoint stiffness matrix. The K_{xx} component (force versus displacement along the direction of instability) increased with the size of the instability, with no change in the K_{yy} component (force versus displacement along the direction of motion) (figure 8.5B). Thus, the increase in stiffness occurred only along the direction to counteract the added instability. Moreover, the net stiffness (stiffness after adaptation + the stiffness of the imposed

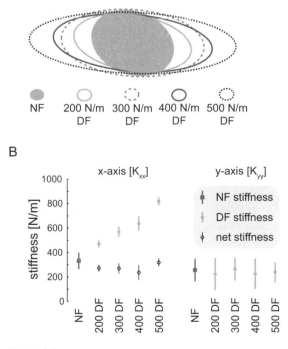

Figure 8.5
Adaptation to instability of different magnitudes. (A) Stiffness ellipses after adaptation to four DF instability levels and the original NF (shaded gray ellipse). (B) Mean and standard deviations of the experimental stiffness along the *x*-axis (K_{xx}) and *y*-axis (K_{yy}) for each level of instability. The net stiffness is the difference between the learned stiffness and the force field strength. Results based on the data of Franklin et al. (2004) and reprinted with permission from Tee et al. (2010).

instability) was equivalent to the original NF stiffness for all levels of instability (figure 8.5B). As the net stiffness is an estimate of the overall stiffness of the coupled system (the limb and environment), this finding suggests that the impedance was tuned by the sensorimotor control system to exactly compensate for the imposed instability and maintain a consistent stability margin under all conditions.

Directionality

Although there is little doubt that the sensorimotor control system can change endpoint stiffness orientation, musculoskeletal mechanics may constrain the ability to achieve optimal tuning (Hu, Murray & Perreault, 2012). For example, the upper limb can be modeled as a two-joint structure with three pairs of antagonistic muscle groups. Each muscle group is able to alter the endpoint stiffness in a specific direction in the horizontal plane (figure 8.6A). Increasing the endpoint stiffness in other directions requires contributions from more than one muscle group. Because stiffness is always positive, these contributions can never cancel, so the ability to selectively increase endpoint stiffness is reduced when more than one muscle group is activated. Three directionally selective DFs were used to test the directional selectively of the endpoint stiffness (Franklin, Liaw, et al., 2007). The DFs were implemented as

$$\begin{bmatrix} F_x \\ F_y \end{bmatrix} = K_x x \begin{bmatrix} \cos\theta \\ \sin\theta \end{bmatrix} \tag{8.2}$$

where the endpoint forces applied to the subject's hand (F_x, F_y) are determined by the field constant ($K_x > 0$) and the displacement of the subject's hand relative from a straight-line movement to the target $(0, y)$. The angle θ was selected from the set $\{-45°, 0°, 80°\}$ to establish the force field orientation. This arrangement created DFs where the instability was directionally specific and selectively proportional to x-axis movement errors (perpendicular to the direction of motion to the target) (figure 8.6C). The three directions were chosen according to the action of the three muscle groups (figure 8.6A). For example, co-contraction of the single joint shoulder muscles should theoretically increase stiffness to compensate for the −45° DF. Similarly co-contraction of the single-joint elbow muscles or biarticular muscles should compensate for the 80° DF and 0° DF, respectively. After adaptation to each force field, the endpoint stiffness of the limb was measured by applying position-controlled displacements at the midpoint of the movement. As predicted, the stiffness increased primarily in the direction of the instability (figure 8.6B), resulting in changes in both the shape and orientation of the endpoint stiffness ellipse (Franklin, Liaw, et al., 2007).

To compare the alignment of changes in the endpoint stiffness with the orientation of directional instability, the change in resistance to endpoint displacement in different directions (increase relative to the NF) was plotted as a force field (figure 8.6D). The changes

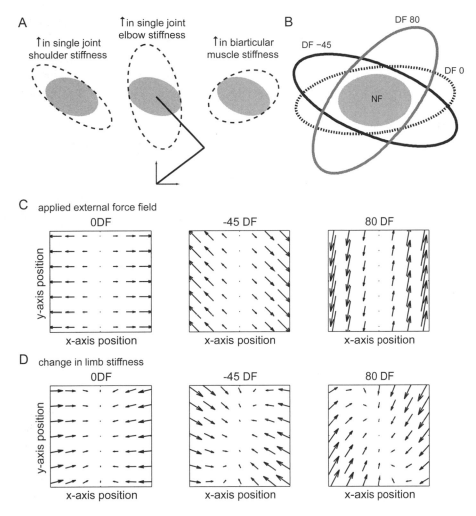

Figure 8.6
Adaptation to instability of different orientations. (A) Simulated changes in endpoint stiffness that would result from increasing the co-contraction of various muscle pairs. The mean experimental NF stiffness ellipse was used for baseline joint stiffness values (shaded gray ellipse). The simulations were performed using the limb posture shown by the stick figure of the arm for pure co-contractions of single-joint shoulder (left), single joint elbow (middle), or biarticular (right) muscles. (B) Experimental endpoint stiffness of the arm in the NF (shaded gray ellipse) and after adaptation to three DFs, each oriented in different directions. (C) The three DFs for a forward reaching motion along the y-axis. (D) Relative change in endpoint stiffness after adaptation to the DF ($\mathbf{K}_{DF} - \mathbf{K}_{NF}$) represented as a force field under the assumption that the endpoint stiffness is independent of perturbation size. Force vectors are plotted as a function of hand displacement. It can be seen that the learned change in endpoint stiffness is appropriately oriented and scaled to oppose the direction of the applied DF. Adapted with permission from Franklin, Liaw, et al. (2007).

in the endpoint force after adaptation were similar in both magnitude and direction to the forces produced by the instability; that is, the force vectors in figure 8.6D are approximately equal and opposite to the force vectors in figure 8.6C. This finding demonstrates that the sensorimotor control system tunes the endpoint stiffness to the environment and provides clear support for the impedance control hypothesis.

Independent Stiffness Control for Different Motions

Our initial studies focused on changes in endpoint stiffness for a single movement direction. However, in many tasks the orientation and degree of instability may vary depending on factors such as movement direction or workspace location. An experiment was performed to determine whether subjects simultaneously optimize the endpoint stiffness for different movement directions or simply select a single stiffness capable of stabilizing all movements required for the task (Kadiallah et al., 2011). Subjects performed reaching movements to two targets separated by 35°, each associated with instability perpendicular to the direction of motion (figure 8.7A). Instability was produced by a DF of the form

$$
\begin{bmatrix} F_\perp \\ F_\| \end{bmatrix} = \begin{bmatrix} K_x x_\perp \\ 0 \end{bmatrix}
\tag{8.3}
$$

where $F_\perp$ and $F_\|$ indicate the force components normal and parallel to the straight line joining the start and end points, respectively, $K_x = -300$ N/m, and $x_\perp$ is the perpendicular deviation of the hand from this straight line. Subjects were not only able to adapt to instability in both movement directions and switch between them from one trial to the next, but the endpoint stiffness for each movement direction was tuned to the instability associated with that motion (figure 8.7B). That is, that there was no difference between the endpoint stiffness when both movements were learned simultaneously and the endpoint stiffness when each movement direction was learned separately. Adaptation to each movement direction occurred through modification of different elements of the joint stiffness matrix, suggesting that the different patterns of co-contraction may have been learned for each movement direction. This finding demonstrates that the sensorimotor system can learn to control the stiffness of the limbs for multiple movements, either through a single state-dependent internal model (as occurs, e.g., in adaptation to stable force fields) or through multiple internal models, each associated with a particular motion.

Stiffness and Noise

Impedance can be used to counteract the effects of motor noise. After adaptation to a DF, the variability of after-effect movements (movements when the DF is unexpectedly removed and replaced by the NF) was less than the initial variability in the NF (Burdet et al., 2001). That is, the increase in endpoint stiffness perpendicular to the direction of movement resulted in after-effect trajectories that were both straighter and less variable

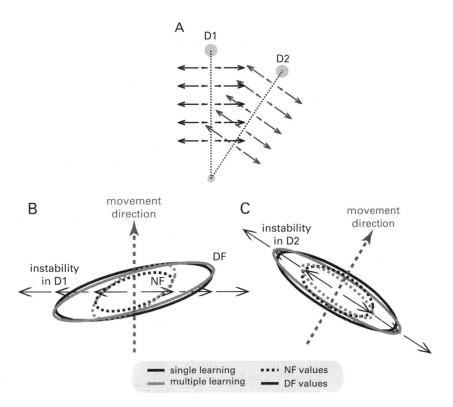

Figure 8.7
Simultaneous adaptation to instability in multiple directions of movement. (A) Subjects performed movements in two directions (D1 and D2), each of which was associated with DF creating instability perpendicular to the direction of movement. These were either performed for each movement direction on separate days (single learning) or with random switching from trial to trial within a single session (multiple learning). (B) Endpoint stiffness in movement direction D1 in the NF (dotted) and DF (solid) after learning separately (single learning) and simultaneously (multiple learning). (C) Endpoint stiffness in movement direction D2 in the NF (dotted) and DF (solid) after learning separately (single learning) and simultaneously (multiple learning). The results indicated that subjects were able to learn the appropriate (optimal) stiffness associated with each direction of movement and switch between them from trial to trial. Adapted with permission from Kadiallah et al. (2011).

(figure 8.3C) than the initial movements in the NF (figure 8.1B). This finding suggests that increased limb stiffness can reduce the effects of motor noise, leading to a decrease in trajectory variability.

This effect may appear counterintuitive: co-contraction requires increased muscle activation, which should create increased motor variability as the result of the signal-dependent nature of motor noise. However, it seems that co-contraction not only limits the increase in noise produced by the increased activation but can also act to reduce the overall variability. Although the noise produced by the activation of one muscle produces variations in the muscular torque acting on the joint, the variations in the joint motion (kinematic variability) depend on the limb impedance—for example, torque applied to a compliant joint produces a greater displacement than torque applied to a stiff joint. Using a detailed neuromuscular model, it has been shown that stiffness increases faster than the noise as muscle activation increases such that an overall reduction in the kinematic variability is achieved (Selen, Beek & van Dieën, 2005) because muscle impedance grows linearly with activation and the independent noise in the multiple muscles grow less than linearly. Therefore, the variability at the joint or limb endpoint level does not necessarily increase linearly with the muscle activity but can be limited and controlled through appropriate tuning of the level of coactivation.

If higher impedance reduces the effect of noise, this relationship could provide a means for improving the accuracy of movements. Several experiments have shown that subjects increase co-contraction (joint stiffness) to adapt to accuracy demands (Gribble et al., 2003; Osu et al., 2004). Conversely, it has been shown that variability in movement endpoint depends inversely on the endpoint stiffness of the limb (Lametti & Ostry, 2010; Lametti, Houle & Ostry, 2007; Selen, Beek & van Dieën, 2006; Selen, van Dieën & Beek, 2006); that is, lower endpoint stiffness or lower co-contraction results in higher variability in the final hand position of reaching movements.

Impedance Control in Static Posture

The ability to manipulate endpoint stiffness under conditions of static posture appears to be limited. Visual feedback of muscle co-contraction (shoulder versus elbow co-contraction) (Gomi & Osu, 1998), visual feedback of the stiffness ellipse (Perreault, Kirsch & Crago, 2002), or instructions to resist force perturbations in specific directions (Darainy et al., 2004) have shown that the ability to selectively modify stiffness in a specific direction appears to be dramatically less when the limb is static than when it is moving. The origin of this difference could potentially be related to mechanics, neural control, or measurement.

Hu, Murray, and Perreault (2012) conducted simulations suggesting that the selective increase in endpoint stiffness found during adaptation to movement in a DF (Burdet et al., 2001) could not be achieved solely through any pattern of muscle activation in a static posture. Although it has been shown that coactivation of biarticular muscles accounts for a large portion of the adaptation to the instability during movements in the DF (Franklin, Burdet et al., 2003), this does not necessarily indicate that the coactivation is entirely responsible for the selective increase in endpoint stiffness. It is questionable whether any pattern of muscle activation that could not produce this selective increase in endpoint stiffness for an arm configuration in a static posture could do so for the same arm configuration during movement. This observation suggests that the measured changes in endpoint stiffness do not simply arise from feedforward changes in the pattern or level of muscle activation.

Another factor that potentially contributes to the different results for posture and motion is the environment used to drive these changes in the endpoint stiffness. The type of environment is critical, as it determines both the modality and relevance of the feedback error signals transmitted to the sensorimotor control system and used to drive the learning of endpoint stiffness. Many of the postural tasks used to study endpoint stiffness involved only visual feedback (Gomi & Osu, 1998; Perreault, Kirsch & Crago, 2002), and the few that produced disturbances that could give rise to proprioceptive feedback involved only simple force perturbations (Darainy et al., 2004; Milner, 2002; Mussa-Ivaldi, Hogan & Bizzi, 1985). Most force perturbations can be counteracted, not just through increased stiffness but also through feedback (reflex) responses or voluntary correction. As the force perturbation is prespecified both in time and magnitude, even delayed feedback corrections can correct the disturbance, as the maximum displacement is limited. On the other hand, instability, as in for example the DF, scales the level of force with the size of kinematic error, which can lead to exceptionally high force levels and large kinematic errors prior to even the fastest feedback corrections. Under these conditions, the optimal solution appears to involve increased stiffness to ensure stabilization.

Several studies have investigated the ability to stabilize the hand in a static posture under conditions of directional instability (Krutky et al., 2010; Milner, 2002), although they did not examine the ability to increase stiffness selectively in the direction of instability. However, endpoint stiffness has been measured in a task that simulated contact instability (Selen, Franklin & Wolpert, 2009). The task involved simulation of the interaction between two circular rigid objects, one associated with the subject's hand and the other fixed in the workspace. The subject was required to apply a target force to the fixed object. By decreasing the diameters of the objects, the instability of the interaction was increased. Measurements of the endpoint stiffness demonstrated modulation with increased instability. However, the endpoint stiffness did not always increase selectively along the tangent to the point of contact of the two objects where the instability would have been greatest. To understand why, a model of the six-muscle, two-link arm complete with signal dependent noise and activation dependent stiffness was simulated in the same task. The simulation demonstrated complex interactions between the stiffness and noise in the multijoint limb. Specifically, it was shown that the optimal change in endpoint stiffness for these object manipulation tasks was not pure increases in endpoint stiffness in the direction of instability. Instead, stiffness increases in other directions, corresponding to the experimental results, were found to produce the required stability with minimal activation. Together, these results demonstrate that both the mechanics (Hu, Murray & Perreault, 2012) and the control (Selen, Franklin & Wolpert, 2009) limit the ability to modulate the endpoint stiffness.

The results demonstrated that within the geometry of a multilink, multimuscle limb, there is a complex interplay between the noise and stiffness. Due to the geometry of the limb, each muscle contributes differentially to the endpoint stiffness, endpoint force, and endpoint noise. The direction of this contribution at the endpoint from each muscle varies with the limb posture. Therefore, these complex interactions can be exploited by the sensorimotor control system to optimize the trade-offs between noise, metabolic cost, stability, and task success. The inclusion of geometry allows the system to manipulate the control strategy, orienting any motor variability at the endpoint in a task irrelevant direction. Thus, the nervous system modulates the limb stiffness in an optimal manner, not increasing the stiffness purely in the direction of instability but in the direction that optimizes with the change in motor noise (Selen, Franklin & Wolpert, 2009).

8.3   Feedforward and Feedback Components of Impedance Control

Adaptation through Co-contraction

As discussed earlier, the analysis conducted by Hu, Murray, and Perreault (2012) suggests that no pattern of feedforward muscle co-contraction could completely account for the selective increase in stiffness measured during movement after adaptation to a DF (Burdet et al., 2001). However, measurement used to estimate the endpoint stiffness was based on the average restoring force over a 140 to 200 ms interval during 300 ms position-controlled displacements (Burdet et al., 2000), which includes contributions arising from both intrinsic muscle stiffness and neural feedback. Monosynaptic and longer latency reflex responses as well as triggered or early voluntary responses could contribute to the measured forces during this interval.

There is no doubt that muscle co-contraction increases during stiffness adaptation, which represents a feedforward mechanism. For example, during adaptation to a DF, activity increases in multiple antagonistic pairs of muscles in advance of movement onset. However, there is also an increase in muscle activity during movement that could arise from feedback responses to correct errors. To test this second possibility, random trials in which only the NF was presented (catch trials in which the possibility of error is minimized and no perturbations occur), were introduced within blocks of DF trials. Thus, as subjects expect the DF, the feedforward muscle activity remains unchanged and feedback responses induced by kinematic errors are eliminated (Franklin, Burdet, et al., 2003). The muscle activity during the catch trials did not differ significantly from the muscle activity during the DF trials (figure 8.8A), indicating that most of the increase in co-contraction was planned in a feedforward manner.

A

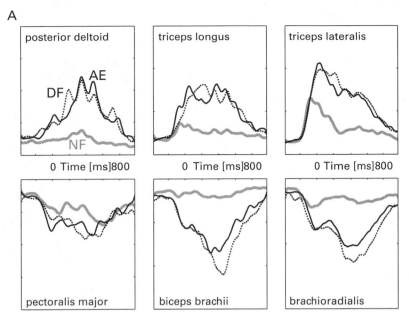

B

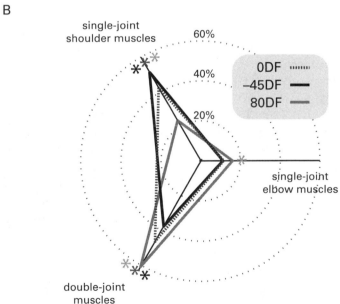

Figure 8.8
Learning feedforward coactivation to compensate for instability. (A) Muscle coactivation is feedforward or predictive in nature. EMG recorded during NF movements prior to learning (thick gray lines), after adaptation to the DF (solid black lines) and during unexpected NF trials within a block of DF trials after adaptation (after-effect trials, AE: dotted lines). The muscle activity in the after-effect trials was almost identical to that in the DF despite the absence of the force field and any errors in the movement, demonstrating that the increase in coactivation was generated predictively rather than through feedback mechanisms. Adapted with permission from Franklin, Osu et al. (2003). (B) Muscle coactivation is appropriately modulated for the direction of instability: unique changes in EMG during movements for movements in three DFs. For each force field, the increase in muscle pair activity relative to NF activity has been plotted as a percentage such that the total change in activity for the three muscle pairs sums to 100 percent. For each muscle pair, a significant difference for a given force field, compared with the other force fields, is indicated with the asterisk. Adapted with permission from Franklin, Liaw, et al. (2007).

The role of co-contraction in stiffness adaptation is also evident from the specificity of changes in muscle activity after adaptation to different orientations of an unstable force field (Franklin, Liaw, et al., 2007). The relative changes in activity of three antagonistic muscle pairs (shoulder, elbow, and biarticular muscles) that contribute to different orientations of the endpoint stiffness ellipse (see figure 8.6A) were quantified for three directions of instability. These changes were tuned to the instability (figure 8.8B); that is, for instability at −45°, activity of shoulder joint muscles increased relatively more than elbow and biarticular muscles, whereas for instability at 80° the reverse occurred. Moreover, the relative changes in co-contraction appeared to be modulated with changes in the configuration of the arm during the motion. For example, as the elbow extends, the ability of single-joint elbow muscles to contribute to stiffness in the direction of instability decreases. This effect was reflected by a decrease in elbow muscle activity and increase in shoulder muscle activity as the elbow extended, suggesting that the impedance controller learns the state-dependent pattern of muscle activity to produce appropriate endpoint stability throughout the movement, although this may be more critical near the end than at the beginning of the movement (Milner, Lai & Hodgson, 2009).

Adaptation through Feedback Tuning
The study investigating stiffness adaptation to directional instability also provided evidence for feedback adaptation (Franklin, Liaw, et al. 2007). Specifically, the off-diagonal terms of the stiffness matrix (K_{xy} and K_{yx}) were found to be significantly different from one another after adaptation to the 80° DF. As intrinsic joint stiffness is essentially conservative in nature (Mussa-Ivaldi, Hogan & Bizzi, 1985), these differences likely arise from feedback asymmetry. Endpoint stiffness could be tuned by selectively increasing the gain of length-dependent feedback to specific muscles. This change would likely incur a lower metabolic cost than tuning stiffness by anticipatory muscle co-contraction but induce a longer response time. Several studies have demonstrated that the reflex gains can change with task stability. As the task changes from a force control task (in which the stability is produced by the environment) to a position control task (in which the stability has to be produced by the subject) or from a stiff to compliant environment, the feedback gain increases (Akazawa, Milner & Stein, 1983; Doemges & Rack, 1992; Perreault et al., 2008). These changes in the reflex gain were shown to occur in multiple muscles at both elbow and shoulder joints (Perreault et al., 2008). It has also been shown that feedback gains are largest in directions where the passive stiffness of the arm is lowest (Krutky et al., 2010). That is, the sensorimotor control system is sensitive to the properties of the limb: when the inherent stiffness of the limb generates enough stability, the feedback gains are smaller.

Reflex contributions to stability have also been demonstrated during ball catching (Lacquaniti & Maioli, 1987, 1989a). When the time of impact could be predicted, stretch reflex gain increased just prior to the impact of the ball (Lacquaniti, Carrozzo & Borghese, 1993).

The increased feedback responses at the time of impact occurred in both stretched and shortened muscles suggesting that the feedback responses are tuned to increase joint stiffness through feedback-initiated coactivation. Moreover, approximately 100 ms prior to the impact of the ball, the muscles of the wrist and elbow were coactivated in a feedforward manner to stiffen the joints against the impact. Therefore, the sensorimotor control system tunes both the feedforward and feedback responses to these predictable disturbances. When subjects were prevented from predicting the time of ball impact (by removing the visual feedback), the predictive feedforward coactivation was absent, but feedback-based coactivation responses occurred at both reflex and voluntary latencies (Lacquaniti & Maioli, 1989b).

Balance between Anticipatory Co-contraction and Position-Dependent Feedback
The main difference between increasing joint stiffness by anticipatory co-contraction and by position-dependent feedback is the balance between immediacy of response and metabolic cost. Anticipatory co-contraction requires a sustained level of muscle activation to set and maintain joint stiffness in anticipation of an unpredictable disturbance, which can involve a high level of energy expenditure depending on the level and duration of the co-contraction. On the other hand, it provides instantaneous resistance to perturbations. In contrast, increasing position-dependent feedback gain likely involves much less energy expenditure, but resistance to a perturbation occurs after significant delays. Although short latency (monosynaptic) stretch reflex responses may begin to produce resistive force at delays as low as 50 ms for proximal muscles, increased gain of these responses is generally associated with higher levels of background muscle activation, which would tend to offset any energy saving (Pruszynski et al., 2009). The longer latency stretch reflexes (which are adaptable in terms of gain and organization) generally cannot produce resistive force until at least 100 ms after the onset of the perturbation. Therefore, it has been suggested that the time for recovery from displacement produced by the instability determines the mechanism for generating stability (Morasso, 2011). The time for recovery from displacement of the body center of mass while standing is relatively long (approximately 900 ms) because of the body's large moment of inertia allowing position-dependent feedback to produce resistive force in time to recover balance. However, object manipulation (Selen, Franklin & Wolpert, 2009) or DFs (Burdet et al., 2001; Franklin, Liaw, et al., 2007) can create situations in which there is relatively little time for recovery (as little as 100 ms), making anticipatory co-contraction and impedance control more effective for maintaining stability.

Impedance Changes during Learning
Adaptation of the multijoint arm to viscous curl force fields induces co-contraction early in learning. This early increase was originally termed "wasted contraction," as it did not contribute to the final adaptation to the force field (Thoroughman & Shadmehr, 1999).

Similarly, large increases in co-contraction were found during initial adaptation to both a stable (velocity-dependent) force field and an unstable (divergent) force field (Franklin, Osu et al., 2003). However, in the stable force field, once the disturbance was counteracted (no kinematic error), the coactivation was gradually reduced to the level in the NF (figure 8.9A), leaving higher activation only in muscles that directly compensated for the force field. In the unstable force field, the initial co-contraction also decayed (figure 8.9B) but in a selective manner such that the final level produced sufficient endpoint stiffness for stability (Franklin, Burdet, et al., 2003; Franklin, Osu, et al., 2003). This finding suggests that greater limb stiffness early in learning is reduced to an optimal level through practice (Franklin et al., 2004). Because disturbances to movement in both stable and unstable environments initially elicit increases in coactivation, it is likely that this strategy represents a functional response to any error (Franklin, Osu, et al., 2003; Milner & Franklin, 2005; Osu et al., 2002). An increase in co-contraction increases the stiffness of the limb (increasing stability) but also reduces the error produced by a disturbance, reducing the deviation from the desired trajectory. When the disturbing forces depend on the state of the limb (the kinematics), then the faster the limb is returned to the "desired" or "optimal" trajectory, the sooner the force experienced by the limb will match the force required to achieve the desired trajectory, speeding up the learning process.

Force Changes during Learning

The sensorimotor control system adapts to stable and predictable forces by changing the net force, gradually adapting to the external dynamics. This adaptation corresponds to iterative learning, in which kinematic error is used to modify the feedforward command from one trial to the next so that error converges to an acceptable value (figure 8.9C). The CNS may attempt to predict the force required to oppose a disturbance even when the disturbance varies in an unpredictable manner. Thus, modulation of the applied force is observed during adaptation to a DF, which can lead to alternation in the direction of kinematic error (Osu et al., 2003) and joint torque (Franklin, Osu, et al., 2003) from one trial to the next. The first trial may deviate to the left or right of a straight line to the target (figure 8.9D). In a DF, the instability of the interaction pushes the hand farther in the direction of that initial deviation. If a corrective force is produced in the direction opposite to the previous error on the subsequent trial, the movement will deviate in the direction of the corrective force and a large error will occur as a result of the amplifying effect of the instability. If this strategy continues, error on successive trials will alternate from one side to the other. In a DF, such a control strategy would not lead to compensation for the disturbance because the error would initially grow and never decrease (Burdet et al., 2006). However, these alternating error patterns gradually attenuate as stiffness produced through muscle co-contraction stabilizes the interaction.

The initial learning strategy in stable or unstable environments appears to involve adaptation of both force and stiffness (Franklin, Osu, et al., 2003). Increasing stiffness during

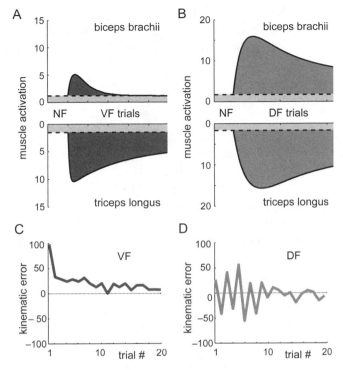

Figure 8.9
Similarities in adaptation to both stable and unstable dynamics. (A) Changes in muscle activation during adaptation to stable dynamics. Initial trials in the novel dynamics elicit large increases in co-contraction of antagonist muscle pairs. As learning progresses, the co-contraction reduces resulting in only changes in muscle activation that produce the required joint torque compensation. (B) Changes in muscle activation during adaptation to unstable dynamics. Initial trials in the novel dynamics elicit large increases in co-contraction of antagonist muscle pairs. As learning progresses, the co-contraction reduces, resulting in changes in coactivation that tune the endpoint stiffness to the environment. (C) Mean integrated kinematic error across subjects during the first twenty trials of adaptation to the stable VF. The error is reduced steadily because the co-contraction limits the disturbance and the appropriate compensation is learned. (D) Mean integrated kinematic error across subjects during the first twenty trials of adaptation to the unstable DF. The error initially oscillates from trial to trial, indicating a similar adaptation strategy as to stable dynamics. As the co-contraction increases and starts to stabilize the system, the oscillations reduce. Figure based on results from Franklin, Osu, et al. (2003) and Osu et al. (2003).

the early phase of learning reduces error resulting in faster convergence to the desired trajectory. Moreover, if adaptation occurs through the tuning of state-dependent basis functions (Gonzalez Castro, Monsen & Smith, 2011), then faster learning will result when the desired trajectory is achieved sooner.

8.4   Computational Algorithm for Motor Adaptation

Computational algorithms for adaptation to stable environmental dynamics can operate in joint space because they must be capable of learning the net joint torques needed to compensate for disturbances (chapters 6 and 7). However, to adapt to instability an algorithm must also be capable of learning appropriate endpoint limb impedance. This capability may be difficult to achieve through joint space mechanisms, particularly when biarticular muscles are involved. On the other hand, it could be straightforward in muscle space. Consider the simple case of two muscles producing opposing torques at a joint. Assuming that movement error is represented by muscle length, an error that stretches a muscle (relative to its length along the desired trajectory) on one trial could be compensated for by an increase in force in that muscle on the next trial. With this mechanism, errors to the left and right of the desired trajectory would lead to a pattern of alternating increases in activation of the two muscles, that is, to coactivation. As muscle impedance increases with activation (Gomi & Osu, 1998; Hunter & Kearney, 1982), this increase in coactivation would increase joint impedance and stabilize the motion. The same mechanism could be extended to multiple joints, including joints crossed by biarticular muscles.

There is experimental evidence though that impedance adaptation involves increases in activation of shortened as well as lengthened muscles (Franklin, Osu et al., 2003). This can be clearly seen during adaptation to the VF (figure 8.9A), where activity increases in the biceps brachii, which was shortened by the trajectory deviation. However, muscle activity cannot increase indefinitely; both under stable and unstable conditions, muscle activity gradually decreases after an initial increase in all muscles (Franklin, Osu, et al., 2003) (figure 8.9A,B).

Together, these experimental results led to the formulation of the following *principles of motor adaptation*. Motor commands to perform a desired action are composed of a feedforward command and a feedback command. Based on empirical evidence, three principles for adaptation of the feedforward command have been proposed (Franklin et al., 2008):

• Errors that stretch a muscle relative to its length along the desired trajectory lead to an increment in feedforward activity in that muscle on the subsequent movement that is proportional to the size of the error.

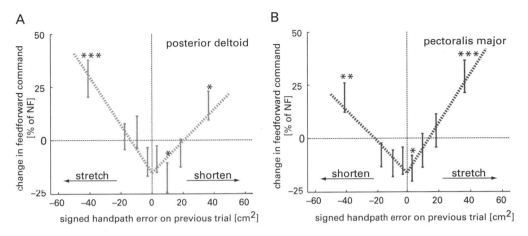

Figure 8.10

Changes in feedforward motor command during adaptation to novel dynamics. The feedforward muscle activity was calculated during initial learning trials in force field adaptation experiments across several force fields (stable and unstable) and plotted against the kinematic error on previous trials. (*A*) Change in feedforward activity of the posterior deltoid (mean ± standard error of the mean) plotted against the handpath error on the previous trial. Asterisks indicate significant differences from zero. (*B*) Change in the feedforward activation of the pectoralis major. Both muscles showed similar responses: if the muscle either lengthened or shortened on the previous trial, the feedforward command increased on the subsequent trial. However, if the signed handpath error was close to zero on the previous trial, then the feedforward command of both muscles was reduced. Reproduced with permission from Franklin et al. (2008).

- Errors that shorten a muscle relative to its length along the desired trajectory also lead to an increment in feedforward activity, but the proportionality constant is lower than for errors that stretch the muscle.

- Muscle activation is reduced as error decreases with practice.

These principles were verified for learning in stable and unstable force fields by relating the trial-by-trial changes in muscle activity (EMG) to trajectory error (Franklin et al., 2008). Muscle activity was shown to increase in muscles that were stretched or shortened by large trajectory errors (figure 8.10). The increase per unit error was smaller when the error shortened the muscle than when it stretched the muscle, creating a change in net joint torque that reduced the error. However, when the error was close to zero, the muscle activity decreased slightly on the subsequent trial.

The three principles were incorporated into a computational model of adaptation (Franklin et al., 2008; Tee et al., 2010). We assume that each of the motor commands $\mathbf{w} \equiv (w_1, \cdots, w_i, \cdots, w_M)^T$ for the M muscles involved in a movement is composed of a feedforward term $\mathbf{u}$ (corresponding to learned dynamics), and a feedback term $\mathbf{v}$ (corresponding to feedback responses from unlearned dynamics):

$$\mathbf{w} = \mathbf{u} + \mathbf{v} \tag{8.4}$$

The feedforward motor command, u_i^k, for each muscle i is updated from trial k to trial $k+1$ by means of the following learning algorithm under the constraint that it must remain positive:

$$u_i^{k+1}(t) \equiv \left[u_i^k(t) + \Delta u_i^k(t+\varsigma)\right]_+, \quad [\,\cdot\,]_+ \equiv \max\{\cdot, 0\}$$
$$\Delta u_i^k(t) = \alpha\left[\varepsilon_i^k(t)\right]_+ + \beta\left[-\varepsilon_i^k(t)\right]_+ - \gamma, \quad \alpha > \beta > 0, \quad \gamma > 0 \qquad (8.5)$$
$$\varepsilon_i^k(t) = e_i^k(t) + \delta \dot{e}_i^k(t), \quad \delta > 0$$

where $e_i^k(t)$ is the stretch/shortening in muscle i at time t, Δu_i^k is shifted forward in time by $\varsigma > 0$, which is equal to the feedback delay, and α, β, and γ are as shown in figure 8.11A. As learning progresses, this algorithm reduces instability, systematic error, and energy expenditure. The current trajectory error specifies the change in feedforward activation in each muscle on the subsequent trial (figure 8.11B,C). This change in activity is shifted forward in time on the subsequent trial in order to deal with neural delays and produce a predictive compensation for the disturbance. These changes in the feedforward motor command lead to a time-varying pattern of muscle activation that achieves the desired movement with minimal muscle activity (figure 8.11D).

Simulation of Adaptation to Stable and Unstable Dynamics

In order to demonstrate the viability and generality of the algorithm, simulations were performed to reproduce the results of different experimental studies (Burdet et al., 2001; Franklin, Liaw et al, 2007; Franklin, Burdet, et al., 2003; Franklin, Osu, et al., 2003; Scheidt et al., 2000). The simulations were carried out using a two-joint six-muscle model of the arm and a learning algorithm that updated the muscle force. The model included rigid body dynamics of the arm segments, activation-dependent intrinsic muscular imped-ance, delayed neural feedback, and signal-dependent motor noise (Franklin et al., 2008; Tee et al., 2010).

The simulation was tuned to produce NF movements similar to those observed in experi-ments in terms of mean trajectory and variability (figure 8.12A). The feedforward command gradually adapted to a VF change in the environmental dynamics, converging monotoni-cally to the NF trajectory (figure 8.12B). If the change in environmental dynamics intro-duced a DF instability (figure 8.12C), the initial trials were perturbed either to the right or the left because of the interaction between the motor variability and the environmental instability similar to human subjects. However, as learning progressed, movements became stable and successfully reached the target. For both the VF and DF, the endpoint stiffness (comprising both intrinsic and reflexive stiffness) changed during learning, resulting in the same characteristics as observed experimentally (Franklin, Burdet, et al., 2003) (figure 8.12D). This change in endpoint stiffness was produced by the combined action of a larger intrinsic stiffness and larger reflexive force, arising from changes to the feedforward command. The simulation (Tee et al., 2010) was also able to reproduce both the scaling

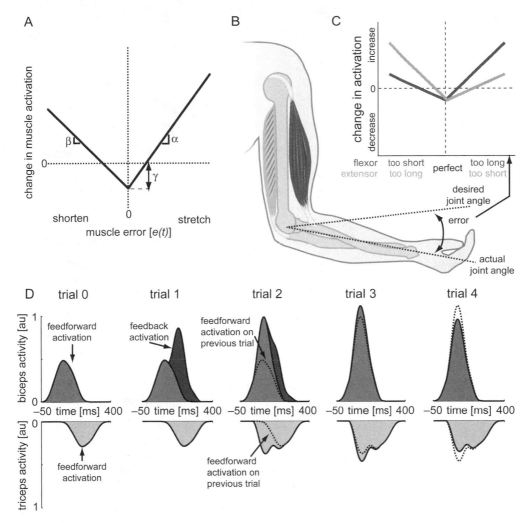

Figure 8.11
Error-based learning algorithm ensures stability and adaptation. (A) The V-shaped learning function. The learning parameters $\alpha > \beta$ determine the slopes of the V-shaped function, whereas γ determines the vertical offset which acts to minimize the activation. (B) For simplicity, the algorithm is explained for a single joint with two opposing muscles: the biceps and triceps. During a movement, the current joint angle is compared to the desired joint angle to give rise to a time-varying sequence of errors. (C) Each error measure is used by the V-shaped update rule to determine the change in muscle activation for the next repetition of the movement. This change in muscle activation is shifted forward in time on the subsequent trial to compensate for the delays. The V-shaped learning rule for each muscle has a different slope depending on whether the error indicates that the muscle is too long or too short at each point in time. (D) Changes in muscle activation patterns during repeated trials. In the steady-state condition, a movement is produced by coordinated feedforward activation of an agonist-antagonist pair (trial 0). The introduction of a disturbance that extends the elbow joint (trial 1) produces a large feedback response in the biceps muscle (darker shading). On the subsequent trial (trial 2), the learning algorithm changes both the biceps and triceps activation pattern (shifted forward in time from the feedback response). Due to this action, a smaller feedback response is produced on this trial. In the next trial (trial 3), the feedforward activation is again increased based on the error on the previous trial such that the disturbance is compensated for perfectly, leading to a reduction in the coactivation on the next trial (trial 4). Figure based on Franklin et al. (2008) and reproduced with permission from Franklin and Wolpert (2011).

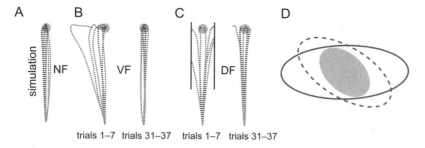

Figure 8.12
The learning algorithm adapts to stable and unstable interactions reproducing experimental data (computational simulations). (A) NF trajectories. (B) Adaptation to the stable VF. (C) Adaptation to the unstable DF. (D) Changes in endpoint stiffness after adaptation. Stiffness ellipses in the NF (filled) and after adaptation to the VF (dashed) and DF (solid) estimated from simulated perturbations after completion of model learning. Reproduced with permission from Franklin et al. (2008).

of the stiffness ellipse with the magnitude of the environmental instability (Franklin et al., 2004) (figure 8.13A,B) and the change in orientation of the stiffness ellipse with the directional rotation of the environmental instability (Franklin, Liaw, et al., 2007) (figure 8.13C). Furthermore, if compensation for a position-dependent instability requires a net force (Osu et al., 2003), the algorithm learns to produce both the required force and change in endpoint stiffness to achieve the desired trajectory (Tee et al., 2010).

Generalization in Motor Learning

When learning a novel task that requires movements in different directions under different environmental dynamics, it is not efficient to represent all commands necessary to accommodate every situation. Therefore, the CNS generalizes learning from previously trained movements, learning to generalize both limb force and impedance according to the environment. The algorithm described previously, able to learn a single movement, was extended to multiple movements (Kadiallah, Franklin & Burdet, 2012) by generalizing the inverse model over the state space $(\mathbf{q}, \dot{\mathbf{q}})$. As the feedforward control signal needs to be able to control arbitrary movements within this state space, it can be coded in a *radial basis neural network* mapping (as done for the learning of torque in chapter 6), defined as follows:

$$\boldsymbol{\mu} = \mathbf{W}\boldsymbol{\Phi}, \quad \boldsymbol{\Phi} = (\phi_1, \phi_2, \cdots, \phi_N)^T, \quad \phi_j(\mathbf{s}_u) = \exp\left[\frac{\|\mathbf{s}_u - \mathbf{s}_j\|^2}{2\sigma_j^2}\right] \tag{8.6}$$

where $\phi_j(\mathbf{s}_u)$ are the nonlinear functions of the hidden layer comprising N neurons.

The input to the radial basis function (RBF) is a vector with components in a state space, such as joint position or both position and velocity $\mathbf{s}_u \equiv (\mathbf{q}_u, \dot{\mathbf{q}}_u)$. Each neuron ϕ_j is centered on a different element $\mathbf{s}_j$ of the state space and has an activation field size defined by the

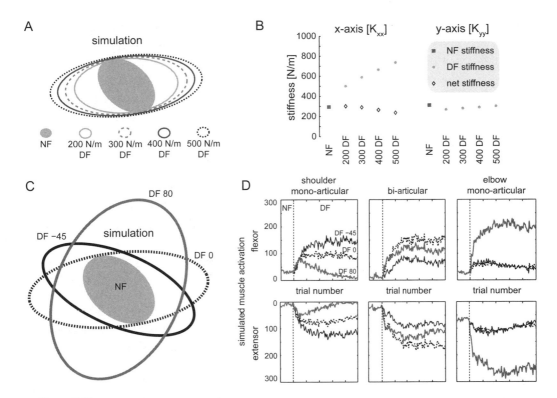

Figure 8.13
The learning algorithm adapts to lateral and directional instability reproducing experimental data (computational simulations). (A) Simulated stiffness ellipses after adaptation to four levels of DF instability and the original NF (shaded gray ellipse). (B) Mean and standard deviations of the simulated stiffness along the x-axis (K_{xx}) and y-axis (K_{yy}) for each level of instability. (C) Simulated endpoint stiffness of the arm in the NF (shaded gray ellipse) and after adaptation to three DFs (see figure 8.6), each oriented in different directions. (D) Changes in simulated muscle activation during learning the three directionally unstable force fields (DF 0, DF –45, DF 80). The learning algorithm is able to learn the appropriate coordination of muscle activity to produce the appropriate endpoint stiffness ellipses and generate the motion. Reproduced with permission from Tee et al. (2010).

deviation σ_j that is set differently for each neuron. The difference between the input vector to the network $\mathbf{s}_u$ and a predefined neuron center $\mathbf{s}_j$ is filtered by a Gaussian function ϕ_j, and the output of the network arises as the product of the weight parameters matrix $\mathbf{W}$ with the vector of outputs from the Gaussian functions.

Similar to that shown in chapter 6, it can be shown that the learning law generalizing the V-shaped adaptive algorithm yields (Kadiallah, Franklin & Burdet, 2012):

$$\mathbf{W}^{k+1}(t) = \left[\mathbf{W}^k(t) + \Delta\mathbf{W}^k(t+\varsigma)\right]_+, \quad \Delta w_{ij}^k = \alpha v_i \phi_j - \gamma$$
$$v_i = \left[\varepsilon_i^k\right]_+ + \chi\left[-\varepsilon_i^k\right]_+, \quad \varepsilon_i^k(t) = e_i^k(t) + \delta\dot{e}_i^k(t), \quad 1 > \chi > 0, \quad \alpha, \gamma, \delta > 0 \tag{8.7}$$

where e is the change in muscle length, k is the trial number, α is the learning factor, and γ is a deactivation constant that reduces superfluous co-contraction (Franklin et al., 2008; Tee et al., 2010). Therefore, the weights of the network are updated after each trial based on the feedback.

The recent experiment of Kadiallah et al. (2011) was simulated in order to test the capability of the model to adapt to instability. In that experiment, subjects performed reaching movements with lateral instability to two targets separated by 35° (figure 8.7A) as described previously. Simulations of hand trajectories and endpoint stiffness during adaptation to a DF in both D1 and D2 directions resulted in similar patterns of adaptation as experimental data (Kadiallah et al., 2011), with the model adapting simultaneously to the instability in the two directions (Kadiallah, Franklin & Burdet, 2012). After learning in the DF, stiffness increased in size for both D1 and D2 movements with the stiffness ellipse elongated in the respective direction of instability. This same model of adaptation was also able to replicate adaptation to stable interactions in multiple experiments, illustrating generalization to different movement trajectories (Conditt, Gandolfo & Mussa-Ivaldi, 1997) and adaptation to force fields of increasing complexity (Thoroughman & Taylor, 2005). Extension of the human-inspired learning algorithm (Franklin et al., 2008) to a robotic system has been implemented, with several demonstrations of its capabilities, as described in chapter 11.

8.5 Summary

The sensorimotor control system is contaminated by noise, arising from stochastic processes such as synaptic transmission, ion channel gating, and motor unit recruitment, as well as history-dependent muscle mechanics. The variability in performance caused by this noise is amplified by any instability in the environment, which can lead to large movement errors.

When the human sensorimotor system is faced with unpredictable or unstable dynamics, it adapts the stiffness of the limbs through co-contraction and changes in position-dependent feedback gain. Adaptations in stiffness are tuned such that the magnitude and direction of the additional resistance are appropriate to compensate for the instability and optimized to maintain net stiffness close to that observed under NF conditions. This constant stability margin may enable the sensorimotor control system to use similar action plans independent of the environment, which could simplify the control process in a modular organization.

When initially exposed to novel environmental dynamics that produce endpoint trajectory error, endpoint stiffness increases and net endpoint force changes through increased co-contraction and relative changes in muscle activation, respectively. Similar changes in muscle activation are found during the initial perturbed trials when adapting to either stable or unstable dynamics.

A computational model of motor adaptation was developed that modifies activation in each muscle independently of the other muscles in response to errors in muscle length. This model specifies feedforward changes in both reciprocal activation and coactivation of muscles and iteratively learns to compensate precisely for forces and instabilities in arbitrary directions. This design produces changes in endpoint stiffness, force, and muscular activation similar to those observed in experiments. With practice, the algorithm learns coordinated control of the redundant muscles, resulting in a behavior that reduces instability, systematic error, and energy.

9 Motion Planning and Online Control

The previous chapters have described the neuromechanical control of the upper limb and its adaptation to modifications in the environment conditions. Feedback and feedforward control and adaptations associated with motor learning and memory have been discussed. However, we have not yet examined how the sensorimotor control system actually plans and implements a movement to perform a desired task, which is the goal of this chapter. Complex actions need to be planned in advance because of the time needed to perform the actions and the large sensory delays. For example, when playing table tennis, it is necessary to infer the path that the ball will follow and plan how to hit it. One can select from different strategies, such as hitting the ball as soon as it touches the table or waiting and hitting it as far away from the table as possible. Furthermore, motion planning should be able to integrate incoming sensory information during movement, for example, to modify the trajectory if the ball deviates from the anticipated path due to miscalculation of the effects of spin or unexpected deflection from the net. To study motion planning and implementation, we start by examining evidence for a plan and the role of such a plan in motion implementation. We then examine the principles governing motion implementation and develop a model of motion control considering the intrinsic sensorimotor noise and integrating sensor signals online.

9.1. Evidence of a Planning Stage

In order to investigate how the nervous system controls relatively complex motion, we performed an experiment in which subjects had to move a stylus on a horizontal display through a sequence of via points as shown in figure 9.1A (Kodl, Ganesh & Burdet, 2011). Two via-point configurations (one of which is shown in figure 9.1A) were presented randomly in three different orientations. Although one may expect that the subjects would always use the same motion patterns, a majority of subjects adopted different strategies in solving this planning problem for different trials performed with the same via-point configuration. Figure 9.1B illustrates three strategies adopted by a typical subject, which are characterized by paths with different curvature signatures. Strategy 1 was the most used

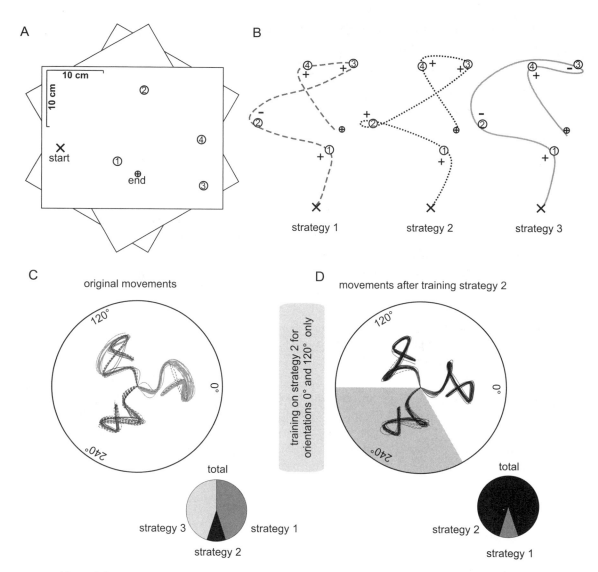

Figure 9.1
Multiple solutions to a via-point task reveal a planning stage. (A) Subjects were asked to make movements through a set of via points presented in three different orientations. (B) Three strategies were used by the subjects, which were characterized by different orders of negative or positive curvature (+/−). (C) The data from one subject's performance. This subject performed the task using three different hand trajectories apparently randomly selected across trials. Later, during a trajectory exploration session, the subject was instructed to perform the hand movements repeatedly along a displayed hand trajectory but only for the 0° and 120° orientations. Following this training, the subject spontaneously displayed (D) an increased tendency to select the trained trajectory not only in the orientations in which it was practiced but also in the unexplored (gray) orientation. Adapted with permission from Kodl, Ganesh, and Burdet (2011).

over all subjects and strategy 3 was least used. The path of strategy 2 always turns in the same direction and thus has the curvature signature [++++] at the via points; strategy 1 is characterized by the signature [+−++] and strategy 3 by [+−−+] (figure 9.1B). Figure 9.1C shows sixty trials performed by the same subject, exhibiting the use of these three different strategies, shown as solid, dashed, and gray lines. These three strategies have significantly different kinematics and require distinct muscle activation patterns (Kodl, Ganesh & Burdet, 2011).

Figure 9.1D shows the trajectories after a period of exposure to strategy 2, in which a template path corresponding to this strategy was presented to the subject on the display for two seconds at the beginning of each trial before movement began. Only via-point configurations in two orientations (0° or 120° for the subject of figure 9.1) were practiced during this phase. Not surprisingly, the proportion of strategy 2 patterns increased for these orientations (figure 9.1D). However, the proportion of strategy 2 patterns also increased in the unexplored orientation (i.e., 240° in figure 9.1D). The bias in the motion pattern selection for the unexplored orientation suggests the existence of a *planning stage* in which the movement profile is determined in extrinsic coordinates before the movement is executed. Note that the movements corresponding to a particular strategy were systematically different for each orientation but shared the same curvature signature (Kodl, Ganesh & Burdet, 2011).

This result corresponds to the traditional hierarchical view of motor control with sensory information or self-intention leading to a motor action via a planning process. In this view, motor planning has been defined as the translation of a general outline of behavior to a concrete motor response through processing in motor pathways (Kandel, Schwarz & Jessell 2000). These pathways have been elucidated over the years from lesion studies, electrophysiology, and brain imaging (figure 2.9). Planning probably starts in the prefrontal areas of the brain before converging to a motor response in the primary motor area following processing of the plan by the posterior parietal, premotor, and supplementary motor areas, which receive inputs from the visual and somatosensory cortices as well as the cerebellum and basal ganglia.

Upper motoneurons in the primary motor cortex exhibit movement related changes in activation just before and during movement. Neurons in areas linked to planning, such as the premotor and supplementary motor cortices, initiate movement related changes in activation that often precede those of the primary motor cortex. However, only the supplementary motor area reflects all of the features of the kinematics to dynamics transformation that would be expected during the planning period prior to movement onset (Xiao, Padoa-Schioppa & Bizzi, 2006; Li, Padoa-Schioppa & Bizzi, 2001). The dorsal premotor cortex reflects them to a degree, and they are essentially absent in the ventral premotor cortex and primary motor cortex, suggesting that the ventral premotor cortex is primarily involved in execution and that the dorsal premotor cortex may be involved in both planning and execution. The picture that emerges is one in which planning and execution represent a mixture of serial and parallel processes. Support for the role of the dorsal premotor cortex in movement planning stems from studies in which it has been shown to encode spatial representation of task goals (Halsband & Passingham, 1985; Jackson & Husain, 1996; Kurata & Hoffman, 1994; Majdandzic et al., 2009; Pesaran, Nelson & Andersen, 2006). However, there are also data that are consistent with a role in motor execution, such as its correlation with reciprocal muscle activation patterns (Haruno et al., 2012).

9.2 Coordinate Transformation

One of the steps involved in planning goal-directed arm movements is the transformation between the coordinate system used to represent the goal and the coordinate system used to control the movement. Planning requires specification of the direction as well as the distance and/or speed at which the hand must move. These factors may be expressed in a visual (gaze-centered) or somatosensory (body-centered) reference frame. The execution of the movement is performed by muscles in a motor reference frame that is body-centered but not necessarily identical to the somatosensory reference frame. Identifying which coordinate system is used in planning human movement has been the subject of recent research. One possibility is that the vector which represents the difference between the current state of the hand and the goal is computed in a body-centered reference frame, as proposed by early psychophysical studies of human movement planning (Flanders, Tillery & Soechting, 1992). However, more recently it has been suggested that this difference vector is represented in a gaze-centered reference frame (Beurze et al., 2010). Neurophysiological studies of neuronal firing in the posterior parietal cortex and dorsal premotor cortex have provided evidence for both gaze-centered and body-centered representations of the target position as well as representation of the difference vector in gaze-centered coordinates (Buneo et al., 2002; Pesaran, Nelson & Andersen, 2006). In the ventral premotor cortex, the predominant representation appears to be gaze-centered, whereas in the primary motor cortex, it is body-centered (Kurata, 2007). However, the distribution of neurons whose firing reflects gaze-centered properties extends into the primary motor cortex, suggesting that movement planning is a highly parallel computational process that may begin in the posterior parietal cortex but includes both the premotor and primary motor cortices, which are primarily involved in motor execution.

The results of initial studies of the activation patterns of motor output neurons in the primary motor cortex were interpreted as encoding the direction of movement. Thus, their activation was characterized in terms of a preferred direction, that is, the movement direction for which their firing rate was highest (Georgopoulos et al., 1982). However, it was shown that the same data could be interpreted as encoding movements in terms of muscle motion or muscle force (Mussa-Ivaldi, 1988). It is unlikely that neurons in the primary motor cortex directly control the force of individual muscles, although there is evidence that they may directly control joint torque (Herter et al., 2007).

9.3   Optimal Movements

What are the features of the movements we carry out according to the plan identified in previous sections? As figure 5.10 exhibits, planar reaching movements where the influence of gravity has been eliminated are generally straight with a bell-shaped velocity profile. To systematically investigate goal-directed arm movements, Flash and Hogan (1985) used a planar manipulandum to record horizontal hand movement. In addition to reaching movements direct to a target, they studied movements with a via point, as shown in figure 9.2A. The smoothness of the via-point movements suggests that both the via point and final target position are included in the movement plan.

Flash and Hogan observed that the via-point movements could be modeled by assuming that they maximize smoothness. In particular, they examined the trajectories corresponding to the minimization of the integral of the square of the *jerk* $d^3\mathbf{x}/dt^3$:

$$V \equiv \int dt \left| \frac{d^3\mathbf{x}}{dt^3} \right|^2 = \int dt \left(\frac{d^3 x_1}{dt^3} \right)^2 + \left(\frac{d^3 x_2}{dt^3} \right)^2 + \left(\frac{d^3 x_3}{dt^3} \right)^2 \tag{9.1}$$

Minimizing this cost function avoids sudden changes in the motion curvature, which would be heavily penalized. The model predicts movement curvature and velocity profiles of horizontal planar movements with a via point similar to those observed experimentally, as can be seen in figure 9.2A.

Although the *maximal smoothness model* is simple and predicts motion patterns for horizontal planar movement similar to experimental observations, it has some important shortcomings. In particular, it does not account for the movement curvature observed when compensating for gravity (Atkeson & Hollerbach, 1985; Papaxanthis, Pozzo & Schieppati, 2003). Natural reaching movements follow a curved path whether executed with or without vision (Day et al., 1998; figure 9.2B,C) and the curvature can depend on the movement direction (Atkeson & Hollerbach, 1985; Papaxanthis, Pozzo & Schieppati, 2003). Furthermore, even horizontal planar arm movements are slightly but systematically curved (Uno, Kawato & Suzuki, 1989; figure 9.2D). Given that it is unclear what functional advantage the maximal smoothness principle could have for the sensorimotor control system and that it is difficult to imagine how the nervous system could compute quantities such as jerk using the noisy sensory signals, alternative optimization models have been proposed.

Several of these models minimize a cost that depends on joint kinematics or muscle activation. For instance, it was proposed that the sensorimotor control system minimizes the integral of the squared derivative of the joint torque (Uno, Kawato & Suzuki, 1989), which corresponds to the minimization of the square of the jerk in intrinsic coordinates (Biess, Flash & Liebermann, 2011). More recently, it was proposed that the sensorimotor control system minimizes the deviation at the target position (Harris & Wolpert, 1998). This paper constitutes a turning point, as for the first time optimization also involved the task to be performed rather than motor variables only. The *minimal endpoint deviation*

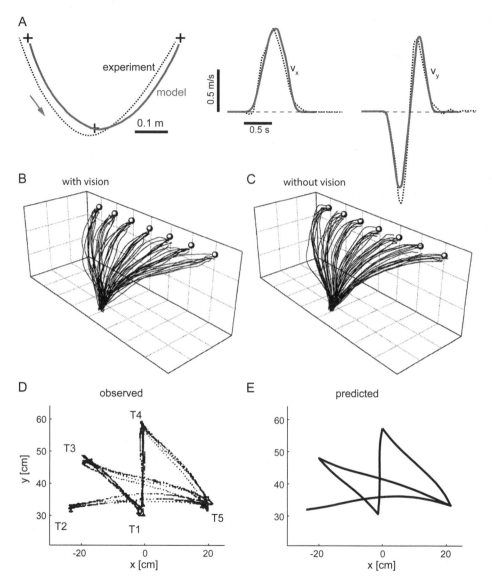

Figure 9.2
(A) Curved goal-directed arm movement with a via point (dotted lines) and prediction of the maximal smooth-ness model (solid lines). (B) Position of the index finger during pointing movements to illuminated targets on a vertical board mounted on a table at which the subject was seated. (C) The same movements as in (B) but made in darkness to the remembered target locations. (D,E) Comparison of actual movement paths (D) and movement paths predicted by the minimum variance model (yielding minimum activation) (E) for horizontal arm movements investigated by Uno, Kawato, and Suzuki (1989). Panel (A) adapted from Flash and Hogan (1985), panels (B) and (C) from Day et al. (1998), and panels (D) and (E) from Harris and Wolpert (1998) with permission.

criterion is practical because it naturally corresponds to the goal and constrains the whole movement: with signal-dependent noise, minimization of endpoint variance corresponds to minimizing effort over the entire movement duration. The cost function is

$$V \equiv \int dt\, \mathbf{u}^T \mathbf{R}^T \mathbf{R}\, \mathbf{u} \tag{9.2}$$

where $\mathbf{R}$ is a matrix weighting activation $\mathbf{u}$ in the different muscles. Figures 9.2D,E show that this model is able to predict the details of movement curvature.

9.4   Task Error and Effort as a Natural Cost Function

As mentioned in section 9.3, the model of Harris and Wolpert (1998) can be interpreted as minimization of either the endpoint deviation or the squared activation. More generally, it is not clear whether the sensorimotor control system is actually minimizing error or effort as noise increases with neural or muscle activity (figure 8.1). In order to systematically investigate how humans consider error and effort in controlling the motor system, we examined how subjects coordinate the force between two fingers (O'Sullivan, Burdet & Diedrichsen, 2009). The experimental setup is shown in figure 9.3A. The subjects had to press against force sensors with the left index/little finger and right little/index finger simultaneously such that the sum of the forces exerted by the two fingers matched a target force level. Theoretically, infinitely many combinations of force exerted by the two fingers could be used to achieve the goal. Signal dependent noise increases linearly with the force applied by each finger (figure 8.1A); however, the rate at which noise increases is different for each finger. Therefore, we could test whether the subjects favored the less noisy finger, shared the force equally between the fingers, or produced a larger force with the stronger finger.

The little finger is weaker than the index and also exhibits greater signal-dependent noise. Therefore, the subjects produced more force with the index finger than with the little finger when the two were paired (figure 9.3D). In contrast, they tended to share the force equally when the pairing involved the same fingers, that is little with little or index with index. To measure the weighting of error and effort, we tested the prediction of minimization of the cost function

$$\int dt\, V(t) \qquad \alpha,\, \gamma,\, \gamma_{MVC} > 0$$

$$V = \alpha\, E\left[(F_l + F_r - F)^2\right] + \gamma\left(u_l^2 + u_r^2\right) + \gamma_{MVC}\left(\left(\frac{u_l}{\text{MVC}_l}\right)^2 + \left(\frac{u_r}{\text{MVC}_r}\right)^2\right) \tag{9.3}$$

where all the variables depend on time t, and the subscripts l and r correspond to the fingers of the left and right hand, respectively. The mean square force error $E\left[(F_l + F_r - F)^2\right]$ penalizes the noisy finger, the second term or effort tends to equalize the forces u_l and u_r

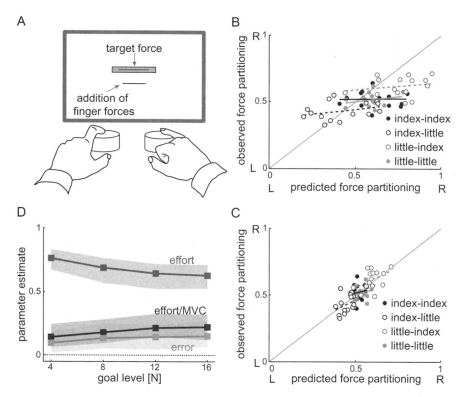

Figure 9.3
Experiment to investigate how two fingers with different noise characteristics share the effort when matching a force target. (A) The experimental setup. (B) and (D) show how the pairs of left and right index/little fingers are used by the subjects if (B) only the deviation is minimized and if (D) effort is also considered. (C) details how error and effort are weighted for the fit shown in (D). Adapted with permission from O'Sullivan, Burdet, and Diedrichsen (2009).

between the fingers, and the third term represents relative effort, normalized by the *maximal voluntary contraction* (MVC).

The parameter values yielding the best fit, shown in figure 9.3C, are plotted in figure 9.3D. These values suggest that although error was considered by the sensorimotor control system, effort was the main determinant of how the force was shared between the fingers. In contrast, considering only the deviation resulted in poor reconstruction of the force (figure 9.3B).

These results are consistent with the principles of motor learning described in chapter 8, according to which error and effort are minimized during practice. This finding supports error and effort as a *natural cost function* for motor behaviors. However, it does not appear that error alone is minimized because the variability in trajectories of NF movements is greater than the achievable minimum, given that variability could be further reduced by first training in a DF (figure 8.3C). As lower variability could be achieved by co-contracting specific muscles—that is, increasing overall muscle activation—this observation suggests that the sensorimotor control system also cares about metabolic cost and relaxes muscles as much as possible while still achieving the task goal.

9.5 Sensor-Based Motion Control

The models described thus far determine the force or activation necessary to achieve a motor task but do not consider how sensory feedback influences the movement. The influence of visual feedback on movement has been investigated by asking subjects to reach to a target as fast as possible with a target that unexpectedly jumps to a new location at various delays with respect to the onset of movement (Georgopoulos, Kalaska & Massey, 1981; Soechting & Lacquaniti, 1983; van Sonderen, Denier van der Gon & Gielen 1988; Paulignan et al., 1991; Flash & Henis, 1995). Figure 9.4A,B shows data obtained with a macaque monkey during such an experiment (Georgopoulos, Kalaska & Massey 1981). We see in figure 9.4B how the movement starts toward the first target and is smoothly redirected after the target jumps. Figure 9.4A shows how the gaze position jumps to the first and then to the second target but does not follow the hand movement. This suggests that the eyes locate the target, enabling the sensorimotor control system to plan and execute the arm movement.

The results of this *double-step target paradigm* showed that the delay between the instant at which the target position jumps and the moment the first change in the trajectory is detected does not depend critically on when the target jumps and requires essentially the same amount of time as that needed to initiate a movement when only one target appears. This means that the visual information is processed immediately when a change occurs, with a constant *reaction time*. This reaction time was estimated to be between 100 and 250 ms based on the first trajectory modification. By moving the location of the target during the saccade to the initial target, Prablanc, Pélisson & Goodale (1986) further showed that subjects modified the movement automatically despite the fact that they did not know that the target location had changed.

A simple mathematical theory to deal with such reaction to incoming sensory information is *linear optimal control*. We will apply this theory to model the hand movement in the experiment. Let us first assume that the hand position is expressed as a linear system in discrete time $k\Delta t$, where Δt is a small time interval:

$$\mathbf{z}_{k+1} = \mathbf{A}_k \mathbf{z}_k + \mathbf{B}_k \mathbf{u}_k \tag{9.4}$$

In this equation, $\mathbf{A}_k$ and $\mathbf{B}_k$ are (time-dependent or -independent) matrices and $\mathbf{z}_k$ is the state vector representing the hand position. For example, a simple model of the arm dynamics can be developed by assuming that a point mass m is moved along the x-axis by a muscle producing a force μ, thus

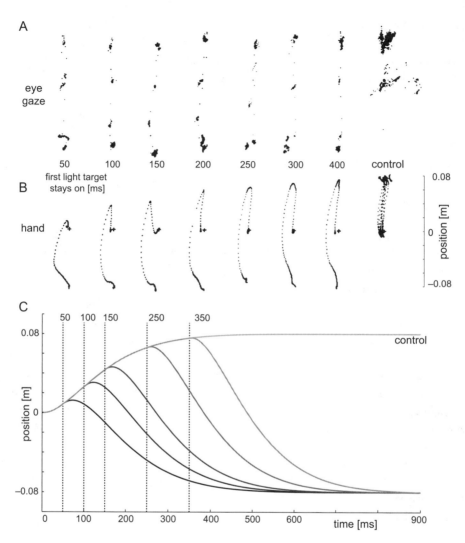

Figure 9.4
Reaction to change in target location during reaching. (A) Eye positions and (B) movements of the hand of a macaque reaching to targets. (C) Simulation of the hand position during target jumps at different delays with respect to movement onset, using a simple model of hand control based on the linear quadratic regular described in section 9.5. Note that the eye gaze jumps between the targets (in advance to the hand) but does not track the movement. Panels (A) and (B) reproduced with permission from Georgopoulos, Kalaska, and Massey (1981).

$$\mu = m\ddot{x} \tag{9.5}$$

We describe the muscle internal dynamics as Winter (2009) does:

$$\varsigma_\mu \dot{\mu}(t) + \mu(t) = \vartheta(t)$$
$$\varsigma_\vartheta \dot{\vartheta}(t) + \vartheta(t) = u(t) \tag{9.6}$$

where $u(t)$ is the motor command and $\varsigma_\mu > 0$, $\varsigma_\vartheta > 0$ are time constants (in seconds) of the muscle mechanics. Using the discrete time approximation $t \to k\Delta t$:

$$\dot{x}(t) \to \frac{x_{k+1} - x_k}{\Delta t}, \; \ddot{x}(t) \to \frac{\dot{x}_{k+1} - \dot{x}_k}{\Delta t}, \; \dot{\mu}(t) \to \frac{\mu_{k+1} - \mu_k}{\Delta t}, \; \dot{\vartheta}(t) \to \frac{\vartheta_{k+1} - \vartheta_k}{\Delta t}$$

in equations (9.5) and (9.6) yields

$$\mathbf{z}_{k+1} \equiv \begin{bmatrix} x_{k+1} \\ \dot{x}_{k+1} \\ \mu_{k+1} \\ \vartheta_{k+1} \end{bmatrix} = \begin{bmatrix} 1 & \Delta t & 0 & 0 \\ 0 & 1 & \dfrac{\Delta t}{m} & 0 \\ 0 & 0 & 1 - \dfrac{\Delta t}{\varsigma_\mu} & \dfrac{\Delta t}{\varsigma_\mu} \\ 0 & 0 & 0 & 1 - \dfrac{\Delta t}{\varsigma_\vartheta} \end{bmatrix} \begin{bmatrix} x_k \\ \dot{x}_k \\ \mu_k \\ \vartheta_k \end{bmatrix} + \begin{bmatrix} 0 \\ 0 \\ 0 \\ \dfrac{\Delta t}{\varsigma_\vartheta} \end{bmatrix} u_k \tag{9.7}$$

which is in the form of equation (9.4).

Returning to the general linear system of equation (9.4), we wish to determine a series of motor commands $\{\mathbf{u}_k\}$ that brings the hand from 0 to the target position $\mathbf{x}^*$ at time $n\Delta t$ with minimal effort. To do so, we minimize the cost function

$$V = \mathbf{e}_n{}^T \mathbf{Q} \mathbf{e}_n + \sum_{k=0}^{n-1} \mathbf{u}_k{}^T \mathbf{R}_k \mathbf{u}_k, \; \mathbf{e}_n \equiv \mathbf{x}^* - \mathbf{x}_n \tag{9.8}$$

where the matrices $\mathbf{R}_k$ are positive definite and $\mathbf{Q}$ is *positive semidefinite*, that is, $\mathbf{z}^T \mathbf{Q} \mathbf{z} \geq 0 \; \forall \mathbf{z}$. $\mathbf{e}^T{}_n \mathbf{Q} \mathbf{e}_n$ results in the system with the dynamics of equation (9.4) converging to the target, and $\sum_{k=0}^{n-1} \mathbf{u}_k{}^T \mathbf{R}_k \mathbf{u}_k$ regulates the transients during motion. It can be shown (e.g., in Stengel [1994]) that the solution to this minimization problem can be computed iteratively, backward in time, as

$$\mathbf{S}_n = \mathbf{Q}, \; \mathbf{S}_k = \mathbf{Q} + \mathbf{A}^T{}_k (\mathbf{S}_{k+1} - \mathbf{S}_{k+1} \mathbf{B}_k (\mathbf{B}^T{}_k \mathbf{S}_{k+1} \mathbf{B}_k + \mathbf{R}_k)^{-1} \mathbf{B}^T{}_k \mathbf{S}_{k+1}) \mathbf{A}_k,$$
$$\mathbf{u}_k = -\mathbf{L}_k \mathbf{z}_k, \; \mathbf{L}_k = (\mathbf{B}^T{}_k \mathbf{S}_{k+1} \mathbf{B}_k + \mathbf{R}_k)^{-1} \mathbf{B}^T{}_k \mathbf{S}_{k+1} \mathbf{A}_k. \tag{9.9}$$

This *linear quadratic regulator* (LQR) can be used as a simple and powerful tool to simulate how the nervous system controls movements while integrating sensory signals.

Figure 9.4C shows how the LQR can be used to simulate reaching with a change in target position. This simulation assumed an arm with dynamics corresponding to the model of equation (9.7), where $\varsigma_\mu = \varsigma_\vartheta \equiv 0.1$ s are the time constants of the muscle/arm and $m = 0.5$ kg, the mass of the arm. For simplicity, the time to process visual information was neglected and we assumed that the nervous system had accurate knowledge of the current position. The task consisted of moving the hand from $x = 0$ toward $x = 8$ cm with a target jump at time $t = T_0 > 0$ to $x = -8$ cm. The simulation used a total movement time $T = 0.9$ s, a time step of $\Delta t = 1$ ms, $\mathbf{R} \equiv 10^{-11}$ and the diagonal matrix $\mathbf{Q} = \mathrm{diag}(50, 0.5, 0.01, 0)$. As illustrated in figure 9.4C, the trajectories obtained with this simulation exhibit smooth modification shortly after the target jump is detected as in the experiment of Georgopoulos, Kalaska, and Massey (1981). These results illustrate the capability of the LQR to simulate sensor-based control.

9.6 Linear Sensor Fusion

How can the sensorimotor control system combine different sources of sensory information, each delayed and with inherent noise, to control movement? For instance, imagine that your mobile telephone rings when you are sleeping. How can you locate it on the bedside table so as to shut it off? The sound may not originate precisely from the telephone location, as the table may resonate. Furthermore, your vision may be impaired because of the dark and blurred if you are not wearing your glasses. Your nervous system must consider these noisy auditory and visual signals in order to determine in which direction to move. To pick up the telephone, your nervous system must also combine proprioceptive, visual, and/or auditory information in order to ensure that the hand reaches the telephone.

The sensorimotor control system has to combine several sources of sensory information in order to infer an object location and perform purposeful actions. This section introduces a simple model of this *sensor fusion* based on linear systems theory. In chapter 10, we will examine how it is implemented by the nervous system. We consider a linear system in discrete time $k\Delta t$ subjected to *motor noise* η, modeled as additive Gaussian white noise:

$$\mathbf{z}_{k+1} = \mathbf{A}_k \mathbf{z}_k + \mathbf{B}_k \mathbf{u}_k + \eta_k \tag{9.10}$$

where $\mathbf{A}_k$ and $\mathbf{B}_k$ are (time-dependent or time-independent) matrices. For example, in the previous example, this *system equation* may represent the hand movement with intrinsic motor noise. This system is observed from one or several sensors, yielding the *observation equation*

$$\mathbf{y}_{k+1} = \mathbf{C}_k \mathbf{z}_k + \omega_k \tag{9.11}$$

where $\boldsymbol{\omega}_k$ represents the intrinsic *sensor noise*, modeled as additive Gaussian white noise. We would like to find the best estimate $\hat{\mathbf{z}}_k$ of the state given the measurements represented by equation (9.11) at time k, thus minimizing the mean quadratic error $E\left[\|\mathbf{z}_k - \hat{\mathbf{z}}_k\|^2\right]$. $E[\cdot]$ represents the *expected value* of the stochastic function to which it is applied, that is, the average over many trajectories with random noise. It can be shown (Simon, 2006) that the solution to this minimization problem can be computed iteratively as

$$\mathbf{P}_{(k+1|k)} = \mathbf{A}_k \mathbf{P}_k \mathbf{A}_k{}^T + E[\mathbf{z}_k \mathbf{z}_k{}^T], \ \mathbf{K}_{k+1} = \mathbf{P}_{(k+1|k)} \mathbf{C}_k{}^T (\mathbf{C}_k \mathbf{P}_{(k+1|k)} \mathbf{C}_k{}^T + E[\mathbf{y}_k \mathbf{y}_k{}^T])^{-1},$$

$$\mathbf{P}_{k+1} = (\mathbf{1} - \mathbf{K}_{k+1} \mathbf{C}_k) \mathbf{P}_{(k+1|k)}, \ \mathbf{P}_0 = E[\mathbf{z}_0 \mathbf{z}_0{}^T] \tag{9.12}$$

$$\hat{\mathbf{z}}_{k+1} = \mathbf{A}_k \hat{\mathbf{z}}_k + \mathbf{B}_k \mathbf{u}_k + \mathbf{K}_{k+1}(\mathbf{y}_k - \mathbf{C}_k \hat{\mathbf{z}}_k), \ \hat{\mathbf{z}}_0 = E[\mathbf{z}_0]$$

where $E(\mathbf{z}_k \mathbf{z}_k{}^T)$ is the covariance matrix of noise $\boldsymbol{\eta}_k$ and $E(\mathbf{y}_k \mathbf{y}_k{}^T)$ the covariance matrix of noise $\boldsymbol{\omega}_k$. This *linear quadratic estimator* (LQE) or *Kalman filter* is a simple and powerful tool that can perform sensor fusion in different systems. To compensate for a delay, the movement can be integrated using the propagator $\hat{\mathbf{z}}_{k+1} = \mathbf{A}_k \hat{\mathbf{z}}_k + \mathbf{B}_k \mathbf{u}_k$, that is, equation (9.12) with $\mathbf{K}_k = \mathbf{0}$. In the case of an unknown system modeled as a randomly moving object, such as a fly, the previous Kalman filter of equation (9.12) can be used with $\mathbf{B}_k = \mathbf{0}$.

In robotics, the LQE is the standard tool used to combine information from diverse sensors such as cameras, optical encoders, and range sensors to determine the position of a robot. Although the standard Kalman filter is based on a linear system model, it is possible to deal with a nonlinear system (e.g., a rotation) by linearizing it at every time step and then applying the previous linear estimator. This technique is called an *extended Kalman filter*. More powerful nonlinear estimator techniques have been developed, such as the unscented Kalman filter and the particle filter (Simon 2006).

To illustrate how the Kalman filter works, we simulate the estimation of the hand position in a reaching movement using information from both proprioception and vision. These two signals are modeled as two random variables with different means, subjected to Gaussian white noise with standard deviations σ_p and σ_v, respectively. Based on the scenario described previously, we simulate the conditions with glasses (i.e., small σ_v) and without glasses (with larger σ_v). In this case, the visual representation of hand position is considered as being more accurate than the proprioceptive representation. Such a situation could potentially arise in the case of peripheral neuropathy associated with severe diabetes. As one can see in figure 9.5A, when vision is blurred—that is, σ_v is large—the hand position estimate shifts away from that of the visual signal, as the estimator automatically weights its information less. In contrast, when vision is sharp (figure 9.5B), the estimator relies more on vision so the estimated position is closer to that provided by the visual signal.

A powerful mathematical result states that if a linear system that is subjected to additive Gaussian white noise uses a linear sensor process also subjected to additive Gaussian white noise to perform optimal control while minimizing a quadratic cost function, then

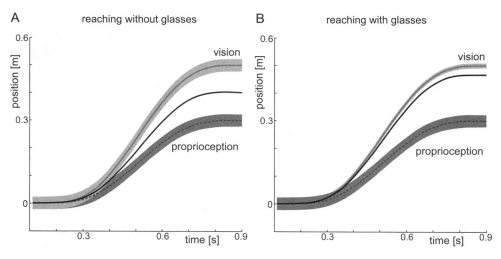

Figure 9.5
Estimation of the hand position for a reaching movement obtained from the fusion of proprioceptive and visual information with glasses (B) and without (A), provided by the linear quadratic estimator presented in section 9.6 under the assumption that vision provides a more accurate estimate than proprioception. We see that without glasses, when the image would be blurred (A), the estimator relies more on the proprioceptive signal even though it is less accurate.

regulation and state estimation are independent. In this case, the solution, called *linear quadratic Gaussian control* (LQG), can be obtained by using the above LQR and LQE together. Considering the discrete case—that is, the system and observation equations (equations [9.10] and [9.11])—the minimization of the cost function equation (9.8) yields equations (9.9) and (9.12). How can we use these results? First, the LQR is computed from the final state backward in time from equation (9.9). Then the LQE of equation (9.12) is applied forward in time along the solution of equation (9.9).

9.7 Stochastic Optimal Control Modeling of the Sensorimotor System

We will now develop a model of sensorimotor control generalizing the mathematical framework described in sections 9.5 and 9.6. This model should determine the force or activation necessary to achieve a motor task while considering both neural feedback and the signal-dependent noise inherent in the generation of movement and in the associated sensory loops. As one can see in figure 8.1, both of these factors are critical even in simple reaching movements. For example, it is apparent in figure 8.1B that these movements have significant variability resulting from sensorimotor noise, as discussed in chapter 8. Furthermore, figure 8.1C demonstrates how visual feedback helps the sensorimotor system to reach the target. The lower final position variability in the presence of visual feedback provides evidence that this feedback is effective because the variability effectively stops increasing as the movement nears the target, whereas without vision the variability continues to increase until movement termination. Figure 8.1C further suggests that there is "minimal" intervention from visual feedback because variability would otherwise be expected to stop increasing earlier along the trajectory, given that the movement duration is 600 ms.

Stochastic optimal control (SOC) (Stengel, 1994; Øksendal, 1998) provides a natural framework to analyze and model how the nervous system controls the musculoskeletal system while integrating sensory feedback and optimally dealing with inherent noise during movement (Todorov & Jordan, 2002). SOC presents several advantages in modeling sensorimotor control. Besides considering the intrinsic noise in the sensory and motor pathways, SOC enables us to express the task goal(s) as additional constraints and avoid other assumptions commonly found in motor control models. For example, it is not necessary to assume a planned trajectory for reaching movements. In fact, an SOC model of human motor control is completely specified by these assumptions: (1) the state of the system is available to the nervous system from the delayed and noisy sensory feedback, (2) the motor plant is subjected to noise that monotonically increases with (neural or muscle) activity, (3) the "effort," generally expressed as the square of the motor command, is minimized together with any task constraints.

A major strength of SOC is that it deals with redundancy in the trajectory, kinematics and muscle mechanics and predicts coordination patterns. The drawback is obviously the potential computational complexity, as all of these aspects of redundancy are dealt with simultaneously. In a redundant system, optimal control determines the task relevant degrees of freedom and does not constrain the other degrees of freedom. Therefore, consistent with actual human movements, SOC does not prevent motion variability but performs online corrections only where and when needed to fulfill the task goal. This *minimum intervention principle* along task-irrelevant dimensions corresponds to the uncontrolled manifold discussed in chapter 5. For example, in reaching to an object, inherent sensorimotor noise results in variability. Optimal control does not constrain irrelevant lateral variability along the movement as long as the target is reached, resulting in nonnegligible lateral variability (figure 8.1).

Mathematically, an SOC model computes the motor command to let the nervous system control the body and reach a task goal while satisfying physical constraints and minimizing a cost. Let the neuromechanical system be characterized by the stochastic differential equation

$$\dot{\mathbf{z}}(t) = \mathbf{f}(\mathbf{z}(t), \mathbf{u}(t), \boldsymbol{\eta}(t), t) \qquad (9.13)$$

where $\mathbf{z}$ is the state space vector fully describing the system at time t (e.g., $\mathbf{z} = (\mathbf{q}, \dot{\mathbf{q}})^T$ in joint space or $\mathbf{z} = (\boldsymbol{\lambda}, \dot{\boldsymbol{\lambda}})^T$ in muscle space), $\mathbf{u}(t) = (u_1, \ldots, u_M)^T$ describes the motor command, $\boldsymbol{\eta}$ the sensori- and motor noise, and $\mathbf{f}(\cdot)$ is the (nonlinear) function describing the evolution of the neuromechanical system. Let physical constraints acting on the system be described by inequalities (e.g., to prevent the movement from entering a certain region)

$$\mathbf{c}(\mathbf{z}(t), \mathbf{u}(t), t) \le 0 \tag{9.14}$$

Then actions are computed to satisfy equations (9.13)–(9.14), which minimize a cost

$$V(\mathbf{u}) = E[h(\mathbf{z}(T_0))] + E\left[\int_0^T g(\mathbf{z}(t), \mathbf{u}(t), t) \, dt\right] \tag{9.15}$$

where h is the *terminal cost* or *task cost*, and g the *running cost*. The SOC problem is to find an admissible control $\mathbf{u}_u(t)$, $0 < t < T$ with trajectories $\mathbf{z}_u(t)$, $0 < t < T$ satisfying the *system equation*, equation (9.13), and subject to the constraints of equation (9.14), such that the expected cost function V is minimized. For example, in the model of reaching movement used to generate figure 9.6A–C, $h = \| \mathbf{z}(t) - \mathbf{z}(T_0) \|^2$ is the terminal cost to reach the target, and the quadratic measure of effort $g(\mathbf{u}) = \mathbf{u}^T \mathbf{R}^T \mathbf{R} \mathbf{u}$ was used as the running cost.

In section 9.6, we pointed out that LQG, formed from LQR and LQE, is the solution to this minimization in the case of a linear system subjected to additive Gaussian white noise, with sensing also affected by additive Gaussian white noise. However, we have demonstrated in chapter 8 that human motor control is characterized by signal-dependent noise increasing monotonically with the motor command, such as

$$\sigma_\eta \propto \mathbf{u} \tag{9.16}$$

In this case, the estimation and regulation are no longer independent, and equations (9.9) and (9.12) have to be modified as described in Todorov and Jordan (2002).

The difference between LQG with additive noise and multiplicative noise is illustrated in figure 9.6A–C. In figure 9.6A, the left panel shows the trajectories of arm reaching movements to a target line, obtained with LQG as described by equations (9.9) and (9.12); the right panel shows the trajectories obtained with signal dependent noise using the algorithm from Todorov and Jordan (2002). Whereas the trials finish close to the line in both cases, the variability during the movement, shown in figure 9.6B, is markedly different in the two cases. In the case of additive noise, the variability increases during the movement by roughly the same amount parallel (y) and perpendicular (x) to the movement direction. Therefore, the ratio of the standard deviations in x and y directions (variability ratio) is close to one throughout the movement (figure 9.6C). In contrast, multiplicative noise during the movement penalizes deviation in the task relevant dimension, y. Therefore, the system minimizes noise in the y direction, particularly near the end of the movement, and does not attempt to correct for lateral deviations that do not affect successful completion of the task. As a consequence, variability in the x direction increases much more as the movement progresses so the variability ratio increases monotonically.

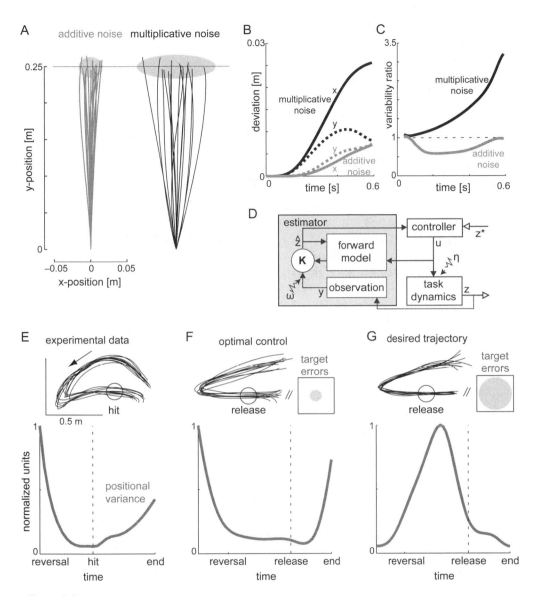

Figure 9.6

Stochastic optimal control modeling of arm movements. (A) Comparison of 15 reaching movements to a line with additive noise (left) and multiplicative noise (right). The principal axes of the shaded ellipse represent the variability (standard deviation) of final positions along the x-and y-axes. (B) Comparison of the mean variability in the hand position in x and y directions for 1,000 trials displayed as a function of time for additive and multiplicative noise as the movement progresses. (C) Ratio of the x and y variability shown in (B). (D) Typical optimal controller with estimator, that is, forward model. (E) Hand position when hitting a ball towards a target with a racquet. Simulated hand position is shown for a similar task in which a ball is thrown towards a target using stochastic optimal control (F) or linear feedback control with a desired trajectory (G). Panel (D) adapted with permission from Rigoux and Guigon (2012) and panels (E,F,G) adapted with permission from Todorov and Jordan (2002).

Figure 9.6D shows the control diagram of a typical SOC model with the estimator (which can be considered as a forward model) and the controller. The characteristics of SOC as a model of human sensorimotor control are further illustrated in figure 9.6E–G, which depicts hitting a ball toward a target. We can observe in figure 9.6E that the subject used a movement reversal strategy, probably to increase the range of motion to achieve sufficient speed prior to hitting the ball with the racquet. The experimental data exhibit nonnegligible variability in consecutive trials. The deviation is minimal at the time the racquet hits the ball (set at peak velocity), which ensures maximal success in hitting the target. Todorov and Jordan (2002) modeled a similar task using a linear stochastic optimal control model (figure 9.6F) and linear feedback control along a desired trajectory (figure 9.6G). They modeled throwing a ball instead of hitting a ball in order to avoid the non-linearity of the ball impact with the racquet.

Figure 9.6F shows that this task can be well modeled using linear SOC with motor noise proportional to the motor command. Movements generated by the model present similar characteristics to actual movements: initial movement backward with direction reversal and large variability overall but minimal variability at the time of ball release, ensuring little variability at the target. As shown in figure 9.6G, linear feedback control along the average trajectory of figure 9.6F can also succeed in hitting the target, albeit with significantly larger variability. Interestingly, in this case the variability over consecutive trials is maximal during the rapid change of force at the movement reversal and larger at the time of releasing the ball as with SOC. Variability becomes negligible at movement termination, but this is not relevant for achieving the best performance. In contrast, optimal control enables the system to focus on the task goal and constraints for maximal success, corresponding to the experimental data.

9.8   Reward-Based Optimal Control

In the previous example, the movement duration was determined beforehand. For example, it was set at 0.7 s to produce the simulation of figure 9.6A,C. Consider the task of reaching for a glass of water or picking up an apple from the table. How long should this movement take? If we consider the effort cost of equation (9.2), it should be as long as possible, because the acceleration decreases in a quadratic fashion with a linear increase in movement duration, as shown by equation (6.13). However, it might be advantageous to satisfy the needs of our body for liquid or energy as quickly as possible using a rapid movement. Similarly, if a monkey has to choose between quickly getting a little food from a nearby location or taking time to get more food from farther away, which option should it select? Stevens et al. (2005) have studied a similar situation and shown how the choice is then modulated by the distance between the two locations.

The trade-off between fast movements to satisfy a need rapidly and the resulting increase in effort can be expressed with the following cost function (Rigoux & Guigon, 2012):

$$V(\mathbf{u}) = \int_0^\infty dt\, e^{-t/\varsigma} g(\mathbf{z}(t), \mathbf{u}(t), t)\,. \tag{9.17}$$

In this equation, ς is the time constant specifying how the *reward* (of achieving the goal, taking the water, the apple, etc.) is "discounted" with increasing task completion time. This cost function, illustrated in figure 9.7, must be maximized in order to maximize the positive reward. Here g is a function which weights a *target state* $\mathbf{z}^*$ where reward is weighted positively and effort negatively, such as

$$g(\mathbf{z}(t), \mathbf{u}(t), t) = \delta(\mathbf{z} - \mathbf{z}^*) - \mathbf{u}^2(t) \tag{9.18}$$

where δ is the delta functional defined by $\int_0^\infty dt\, \delta(\mathbf{z}(t) - \mathbf{z}^*(t)) \equiv 1$. As the time to reach the rewarded state increases, the reward term decreases due to time discounting and the effort term decreases because slow movements are less costly than fast movements. As shown in figure 9.7, the difference between the reward and effort terms defines an *optimal time* T^* for which the objective function is maximum. Using T^*, classical optimal control on a finite time horizon can be applied to derive an optimal control policy. The main interest in this formulation is that T^* is automatically updated during movement if perturbations are applied to the target or to the moving limb.

Note that for simplicity, equation (9.17) corresponds to the deterministic case neglecting sensorimotor noise. This equation can be used not only to determine an optimal solution as will be described shortly but also as a model of learning of behaviors. In fact, equation (9.17) is a formulation of ("reinforcement") learning in continuous time and space (Doya,

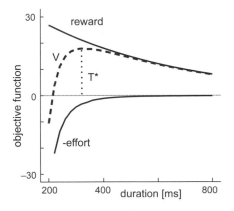

Figure 9.7
Cost function with discounted reward illustrating the trade-off between completing the task as fast as possible and the required increase in effort. Adapted with permission from Rigoux and Guigon (2012).

2000; Todorov, 2007), and optimal movements can be learned by using classical reinforcement techniques (e.g., Shimansky, Kang & He, 2004).

This model can account for the choice of food location in the experiment of Stevens et al. (2005). It has also shed light on several observations related to movement timing. In particular, Gordon et al. (1994) observed a systematic variation in movement duration for reaching in different directions. Simulation of these movements with the model of equation (9.17) predicts this variation in movement duration very well (figure 9.8B), corresponding to the anisotropy of the force needed to move the arm in different directions (dotted line in figure 9.8B). The movement dynamics also induced slightly curved movements (figure 9.8A) that correspond to the experimental results of Gordon et al. (1994).

Another observation of Gordon et al. (1994) is that peak reaching velocity and movement duration increase with movement amplitude, though the general shape of the velocity profile remains roughly the same. This finding was also reported in a number of earlier studies (Bouisset & Lestienne, 1974; Freund & Büdingen, 1978; Wadman et al., 1979; Milner, 1986b). Figure 9.8C,D shows that the model also predicts this result. For reaching with a jump in target location, this model can predict trajectories similar to those observed experimentally, as can be seen by comparing its prediction in figure 9.8E,F with experimental data from, for example, Flash and Henis (1995) or Liu and Todorov (2007). Although the models proposed in both of these papers could also predict these motion patterns, reward-based optimal control can, in addition, predict how the overall movement duration changes with reaction to target change.

9.9 Submotion Sensorimotor Primitives

Optimization can, in principle, be used to coordinate the system of the limbs and muscles. However, practically it may be difficult to learn complex behaviors with a system that has as much redundancy as the human neuromuscular system. On the other hand, biomechanical and neural constraints may impose constraints and thus reduce the redundancy. In particular, the hierarchical organization of the nervous system described in chapter 2 may both constrain and facilitate motor control and learning. A popular idea is that the nervous system uses sensorimotor building blocks or primitives to facilitate motor control and learning. This section presents one kind of such *motion primitive* for planning and executing actions in the form of *submotions*.

Figure 9.9A,B shows the movements of infants attempting to reach an object. In the first five months, the movements are composed of several submovements of similar amplitudes (figure 9.9A). Following a submovement, if there is still distance remaining to the object, a new submovement can be generated to move closer to the target. This behavior undergoes a critical transformation at about six months of age, when the movement becomes smoother and exhibits a large initial velocity peak (figure 9.9B), suggesting that infants become able to coordinate these submovements (von Hofsten, 1991; von Hofsten

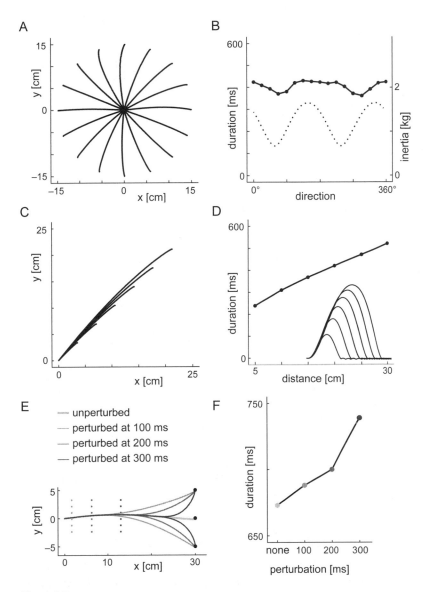

Figure 9.8
Simulation of reaching movements in different directions (A,B), of different amplitudes (C,D) and with target jumps (E,F). The movement paths are shown in panels (A,C,E) and the corresponding movement duration in panels (B,D,F). The inset in (D) indicates how the velocity profile scales with movement amplitude. The dotted line in (B) shows the directional dependence of the inertia, obtained from the resistance to a unit acceleration vector applied in each direction. The movement duration is generally larger for directions with greater inertia. The vertical lines in (E) indicate the positions corresponding to the target jumps. Comparison with figure 3 from Gordon et al. (1994) (A–D) and figure 1 from Liu and Todorov (2007) demonstrates that the optimization of section 9.8 predicts the trajectories and movement time well. Reproduced with permission from Rigoux and Guigon (2012).

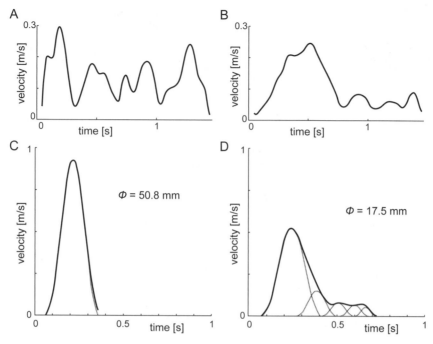

Figure 9.9
Velocity profiles of reaching movements composed of submotions. (A,B) Movements of an infant at nineteen and thirty-one weeks. The control of reaching movements undergoes a critical transformation at about six months from a series of submovements to a smooth, direct movement with a large initial submovement. (C,D) Arm movements of an adult subject placing a peg of diameter 9.5 mm into a hole of diameter 50.8 mm (C) and 17.5 mm (D). Panels (A,B) based on data from von Hofsten (1991) and panels (C,D) reproduced from Milner (1992), with permission.

& Rönnqvist, 1993). This finding also means that infants modify the visuomotor coordination strategy and become able to infer how to generate a continuous movement rather than stopping and starting.

We have shown in chapter 4 that slow arm movements of adults seem to be composed of submovements, as indicated by inflections in the velocity profile and synchronous changes in the motion curvature (figure 4.10). Such submovements were detected in the movements of adults placing a peg into holes of different diameters (Milner & Ijaz, 1990). As can be seen in figure 9.9C,D, the velocity peak decreases and the movement duration increases as the required motion accuracy increases. Small fluctuations in the velocity near the end of the movement can be interpreted as submovements because they correspond to changes in movement direction. These submovements may correspond to visually elicited corrections (Burdet & Milner, 1998). In this view, model-based prediction corresponding to the regularity of submovements could be used to predict where the movement is heading, which then serves as the basis for launching the next submovement.

In the model of Burdet and Milner (1998), each submovement is contaminated by noise that is larger for faster submovements, such that small slow submovements are best used close to the target in order to ensure that the peg enters the hole. The submovements, representing intermittent control, can be used to speed up learning. Using trial and error minimization with a cost function similar to equation (9.17) produced results similar to the experimental data shown in figure 9.9C,D. This model may correspond to the learning by infants in the first six months (figure 9.9A,B). The increase in memory capacity observed at about this age may enable such predictive control (Burdet & Milner, 1998). Note that learning to perform this type of control would not be possible using the error-based learning schemes presented in chapters 7 and 8.

9.10 Repetition versus Optimization in Tasks with Multiple Minima

Many complex actions are characterized by several possible solutions that achieve the goal, similar to the via-point task described in section 9.1. For instance, the orientation of the front wheel of a bicycle can be maintained on a bumpy road using various levels of antagonistic muscle co-contraction (figure 9.10A). However, there exists one particular level of co-contraction that results in an impedance corresponding to front wheel resonance (center peak), which would amplify front wheel shimmy and could ultimately lead to an accident, as one of the authors experienced as a child.

In order to examine how humans deal with a task characterized by multiple minima of error and effort, we performed an experiment with similar resonance dynamics (Ganesh, Haruno, et al., 2010). The subjects were required to keep the wrist within a 3° target window by controlling the co-contraction level of wrist flexors and extensors measured with surface EMG. The wrist was displaced under servo control with a sinusoidal amplitude that was a function of the co-contraction level. The displacement remained within the target window for both low and high levels of co-contraction but was displaced by more than 3° for intermediate levels of co-contraction. This situation is analogous to the one described earlier of steering a bicycle on a bumpy road. The subjects performed 110 s trials during which the oscillation frequency was reduced from 6.5 Hz to 4 Hz in 0.25 Hz steps. Note that these frequencies are too high for the subject to track with reciprocal activation of the wrist flexor and extensor muscles. Trials performed by a typical subject are shown in figure 9.10B.

Approximately half of the subjects started in the low co-contraction region and half in the high-co-contraction region, and all repeated the same protocol. The subject of figure 9.10B started in the high co-contraction region (black solid line); he was then provided with feedback of the co-contraction level and required to start the next three trials with low co-contraction (dashed dark line). In the subsequent three trials, he was free to choose the co-contraction level but remained at the low co-contraction level (dashed light line), which seems logical, as this requires less effort. However, when the subject was again

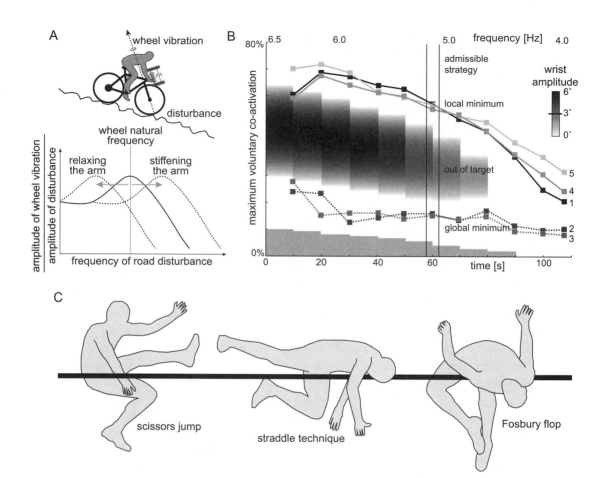

Figure 9.10

Learning to perform tasks with multiple solutions. (A) illustrates the task of steering a bicycle while descending on a bumpy road. The transfer function of a simple second-order system model of this task shows resonance at a frequency which changes with the impedance (stiffness) of the arms. (B) shows the level of wrist muscle co-contraction employed by a subject performing a similar control task in which the wrist must be maintained in a 3° target window while being displaced by a sinusoidal oscillation that varies in amplitude as a function of muscle co-contraction. To succeed—that is, to stay within the white region—the subject can either co-contract the antagonist muscles or relax them. Each line represents the median value from the three consecutive trials performed by the subject. (C) shows three styles of performing the high jump: from left to right, the scissors technique, the straddle, and the Fosbury flop, which illustrate how behaviors evolve in a discontinuous way through imitation and local optimization of a task satisfying solution. Panel (B) adapted from Ganesh, Albu-Schaeffer, et al. (2010) with permission.

required to start at the high co-contraction level for three trials (dark gray solid line), he maintained the high co-contraction level for the subsequent trials although he was free to choose (light gray solid line).

All the subjects exhibited similar behavior over all trials. (1) Their selected co-contraction level did not converge to the global minimum of effort, even though they had experienced this minimum in previous trials. In contrast, they selected the same co-contraction level as on the previous trial. (2) They minimized the effort locally, because they did not maintain the same co-contraction level throughout the trial, but decreased along the boundary of (locally) minimal effort. This latter behavior is consistent with the model of Franklin et al. (2008) described in chapter 8, according to which the behavior would perform a gradient descent minimization of error and effort. In summary, the subjects' behavior was characterized by memory and local effort minimization, so that they would repeat the (possibly suboptimal) strategy employed in previous trials. This behavior may be practical, as it does not require that the sensorimotor control system keep track of all local minima and choose among them.

Though our task was simpler than most tasks of daily living, the observed behavior based on memory and local optimization might explain how humans learn complex tasks, as illustrated in sports by the evolution of the high jump style. After years of adaptation, by the 1960s, the "straddle" style of high jump was considered to be most efficient until self-styled athlete Dick Fosbury introduced the "Fosbury flop," a radically new style that can be shown to be more efficient from the perspective of effort than the straddle (figure 9.10C; Dapena, 1980). This example shows that even after years of exploration, we may not be performing tasks optimally. The large kinematic differences between the straddle and the Fosbury flop gives some indication of how difficult it can be to converge from a suboptimal local minimum (like the straddle) to a more global lower optimum (like that of the Fosbury flop).

This example suggests that instead of attempting global optimization, the sensorimotor control system might rather repeat a strategy that it knows will achieve the goal. The muscle activation can however be locally optimized (Franklin et al., 2008). In this view, knowledge or observation would provide us with the template to imitate, which can subsequently be optimally refined until a more optimal behavior is discovered and adopted. Imitation may help humans achieve gross convergence to globally optimal solutions taught by experienced individuals, while local optimization of newly adopted behaviors would automatically arise with repeated practice (Franklin et al., 2008; Izawa et al., 2008). The competitive advantage of optimal behaviors may explain why the human brain is equipped with explicit networks for imitation learning (Rizolatti & Craighero, 2004).

9.11   Summary and Discussion on How to Learn Complex Behaviors

Generalization of via-point movement patterns to untrained directions demonstrated that arm movements are planned in extrinsic coordinates. The regularities observed in learned movements were modeled using a stochastic optimal control framework, based on inherent motor noise, continuously available sensory signals, and effort minimization, with minimal assumptions not related to the task. This model predicts the trajectory, timing, and large variability over consecutive trials in task irrelevant dimensions, which are observed in simple movements such as reaching, via-point movements, and ball throwing. The mathematical techniques introduced to model continuous control and predict the neuromuscular system state will be used in the next chapter to describe how feedback from multiple (possibly conflicting) signals is integrated to determine the motor control.

Despite its appeal as a framework for modeling sensorimotor control and coordination, optimal control may not be able to deal with complex tasks, nonlinearities of the sensorimotor system, and the multiple strategies available to accomplish most tasks. However, simplifying constraints in the biomechanics or in the neural pathways, such as the submovement hypothesis, may reduce redundancy and facilitate coordination and control.

What are the principles of motor learning that lead to the motor behaviors examined in this chapter? Chapters 7 and 8 described motor adaptation occurring in an automatic and involuntary way by minimizing instability, systematic error, and effort (Franklin et al., 2008). Although this adaptation does not affect the trajectories much, the minimization of activation, together with motion variability in different trials, may slowly lead to trajectory modifications corresponding to the minimization of effort (equation [9.2]), such as observed by Izawa et al. (2008). Furthermore, machine learning studies have provided various techniques to gradually optimize behaviors such as reinforcement learning based on the minimization of an objective function as in equation (9.17) (e.g., Shimansky, Kang & He, 2004) or systematic local trajectory optimization techniques such as PI^2 (Theodorou, Buchli & Schaal, 2010).

However, these techniques are still not adequate to explain how humans can learn complex behaviors such as high jumping or other sports or how robots could learn similar complex behaviors. How can global optimization of complex tasks be achieved? Observations of the performance of a simple task that simulated multiple solutions of a cost function of error and effort suggest that to improve task performance, humans may not need to use a systematic optimization process. Instead, they can simply memorize a task satisfying strategy learned by imitation or imposed by education, which would be automatically optimized locally. This theory would explain how we can progress in a discontinuous way toward an optimal solution and provides a viable solution to improve behaviors in robotic systems.

10   Integration and Control of Sensory Feedback

We perceive the environment, and our state within this environment (e.g., our body position), from our senses—all of which are subjected to inaccuracy, noise, and delays that affect the reliability of the information. The sensorimotor control system must assess the reliability of the sensory information and determine the appropriate action to take at each moment in time. As discussed previously, the noise in sensory feedback can depend on both limb state and location in space. For example, although the position of the arm in space can be sensed through proprioception (including skin, joint, and muscle afferents), the accuracy in this measurement is nonuniform. It is largest in the direction away from the body (in depth) and smallest in the lateral direction (the azimuth) (van Beers, Wolpert & Haggard, 2002). Similarly, the visual feedback of limb position is most accurate in the azimuth and least accurate in the direction of depth. Accuracy depends on the resolution of the sensors and noise in sensor output. Resolution is determined by the amount of change in the stimulus required to produce a change in sensor output, whereas noise represents the variability in sensor output for the same stimulus. Accuracy is also affected by biases in the sensory feedback that change as the limb position varies (van Beers, Sittig & Denier van der Gon, 1998). Thus, estimates of the internal and external states are subject to errors that can vary throughout the state space.

The sensory information is also subject to delays present throughout the sensorimotor system, not only in the transduction (within the receptors) and transmission of the sensation (due to neural conduction and synaptic transmission) but also in the processing that gives rise to the perception and in the neural and mechanical delays in acting upon this information (Franklin & Wolpert, 2011). These delays, on the order of 100 ms, depend on the particular sensory modality (e.g., longer for vision than proprioception) and complexity of processing (e.g., longer for face recognition than motion perception), meaning that we effectively live in the past, with the control systems having access to only out-of-date information about the world and our own body. Moreover, these delays vary across all of the different sources of information.

The combination of sensory noise and delays leads to uncertainty about the current state of the world (uncertainty in both internal and external states). However, uncertainty can also arise from limitations in sensory resolution stemming from sparse receptor density, discrete encoding (i.e., action potential generation), and intermittent sampling of continuous signals, as well as from ambiguity in sensory processing. The sensorimotor system must therefore operate with the inherent noise, delays, and uncertainty in the sensory feedback that it receives. In this chapter, we discuss several computational principles that have been used to explain how the neural system deals with these issues and allow skillful control, namely Bayesian integration, forward models, and adaptive and active feedback.

10.1   Bayesian Statistics

The sensory information we receive is often ambiguous (e.g., the same visual signal on the retina can arise through multiple objects; Ernst & Bülthoff, 2004), leading to uncertainty in the state of the world. In addition, we may be required to act before complete information is obtained, due to delays in both the transmission and processing of sensory feedback. Bayesian statistics is a framework for understanding how the nervous system performs optimal estimation and control in an uncertain world. Bayesian statistics involves the use of probabilistic reasoning to make inferences based on uncertain data, combining uncertain sensory estimates with prior beliefs and combining information from multiple sensory modalities in order to produce optimal estimates.

Bayesian Integration

One manner in which to refine estimates is to integrate over the prior distributions of possible states of the world. This prior distribution reflects that not all possible states of the world are equally likely. For example, if we were required to reach out to touch one of the shapes in figure 10.1, we need to determine whether it is concave or convex. The two-dimensional visual feedback of the shapes is not sufficient to determine the depth because there are an infinite number of three-dimensional shapes that can produce the same two-dimensional image on the back of the retina (Ernst & Bülthoff, 2004). In this case, each of the shapes could potentially range from concave to completely flat to convex, depending on your assumption about the color and direction of light source. However, almost certainly, you perceive concave shapes on the far left column and convex shapes on the far right column. Therefore, your perceptual system has chosen one out of all of the possible shapes as the most likely. However, if the figure is rotated 180°, then the shapes that you perceived as convex will now appear concave and vice versa, because our sensorimotor system combines the visual sensory information with our expectation about from the location of the light source (above us, as, e.g., the sun). This expectation is termed *the prior*. It represents our prior belief about the likely state of events. In this context, with a prior belief that light comes from above, the shapes will be perceived as concave on the left and convex on the right because these shapes are the most likely of all possible shapes to produce the shadows and highlights that are seen in the figure. Bayesian statistics provides a framework for interpreting similar situations, in which the brain decides on an action by relying on a priori information (see Körding & Wolpert, 2006 for a review). It has been used to explain many visual illusions by making assumptions about the prior beliefs of visual objects or direction of illumination (Adams, Graf & Ernst, 2004; Kersten & Yuille, 2003). Bayesian statistics is computed from an optimal combination of probabilities to determine the most likely explanation. Bayes rule is used to calculate the best prediction based on the combination or prior belief and current information. If, for example, we wish to estimate the most likely configuration of our limb joint angles θ based on some sensory feedback s, then we can first compute the *conditional probability* $P(\theta \mid s)$, that is the probability of being in the limb configuration θ given that we have received the sensory information s. By definition

$$P(\theta \mid s) \equiv \frac{P(\theta \cap s)}{P(s)}.$$
(10.1)

To facilitate computation of $P(\theta \cap s)$ we also note that

$$P(s \mid \theta) \equiv \frac{P(\theta \cap s)}{P(\theta)}$$
(10.2)

from which we obtain *Bayes rule:*

$$P(\theta \mid s) = \frac{P(s \mid \theta) \cdot P(\theta)}{P(s)}$$
(10.3)

Bayes rule can be used to calculate the *posterior*, namely, here the estimate of the limb configuration, based on the prior belief and current sensory information. The *likelihood*, $P(s \mid \theta)$, is the probability that we would have received such sensory feedback if we were in that limb configuration, and the prior, $P(\theta)$, is the probability that we could be in that limb configuration. For our purposes, the denominator $P(s)$ is a normalization factor that does not change the relative probabilities. Application of this rule is possible because some body segment configurations are much more common than others (Howard et al., 2009), providing valuable information that can be used to refine our estimate of the current configuration. The posterior (or optimal estimate), obtained by the combination of the sensory information and prior belief, always has lower uncertainty than the estimate from sensory information alone. Therefore, by combining sensory information with prior information, Bayesian estimation acts to reduce the uncertainty in our estimates. The posterior estimate can be considered an optimal estimate, in that it maximizes the probability of the posterior across all possible states. When no prior information exists (a flat prior across all possible states), then maximizing the posterior is identical to maximizing the likelihood (relying purely on sensory feedback in the previous example).

The sensorimotor system has been shown to use Bayesian integration, that is, to combine a representation of the prior and the likelihood in a Bayesian manner to produce an estimate

Figure 10.1
Influence of prior information on our sensory perception. When we view the object, we will normally perceive
alternating columns of concave and convex dimples with the leftmost column concave and the rightmost column
convex. This perception arises from the combination of the sensory information about the shading of the circular
regions with prior evidence that light normally comes from above. Therefore, if one keeps focus on a particular
dimple while the figure is rotated 180°, the previously convex dimple will now look concave and the originally
concave dimples will now appear convex.

of the sensory state. In a study by Körding and Wolpert (2004), subjects were instructed
to make reaching movements while the visual feedback of the hand was displaced laterally
from its actual location based on random draws from a Gaussian distribution with a mean
of 1 cm (and standard deviation of 0.5 cm) (figure 10.2B). Each trial was displaced by a
different amount; on average, the perturbation was 1 cm to the right. Visual feedback was
presented momentarily at the midpoint of the reaching movement to the target (figure
10.2A). Subjects were required to estimate the location of the visual cursor and place it
over the target, thereby compensating for the imposed visual shift. After subjects had been
presented with this distribution of visual shifts for 1,000 trials, all visual feedback was
eliminated to determine whether they had learned the mean of the underlying distribution.
Following removal of visual feedback, subjects shifted their final hand position 0.97 cm
to the left, indicating that they had learned the mean of the distribution. The second part
of the study investigated how this prior information about the distribution would be
weighted with the sensory feedback. The effective variance of the sensory information was
modified by blurring the visual feedback received at the midpoint of the movement (figure
10.2C). If subjects optimally combine the prior information about the distribution with the
current sensory feedback, then as the visual blur increases, the relative weighting placed
on the prior information should increase (figure 10.2D). On the other hand, with perfect
visual feedback, the prior should have almost no influence on the estimate (figure 10.2E),
which is exactly what was found experimentally (figure 10.2F), suggesting that subjects

learned the characteristics of the prior distribution of the discrepancy, had an estimate of the reliability of the visual information so as to estimate the likelihood, and combined these two sources of information in a Bayesian manner. Further support for Bayesian integration in the sensorimotor system has been found in force estimation (Körding, Ku & Wolpert, 2004) and interval timing tasks (Miyazaki, Nozaki & Nakajima, 2005).

Multisensory Integration

For many tasks, state estimation need not rely on a single modality of information. Different sensory modalities often sample information about the same state of our bodies (e.g., visual and proprioceptive estimates of hand position) or the same state of the external world (e.g., visual and auditory estimates of the position of a bird). Most of the time, the different sensory modalities provide complementary information. However, experimentally they can be made to conflict in order to investigate the computational mechanism that determines the combination of sensory information. For example, by mismatching the visual and auditory information of a person speaking, a percept can be obtained that is intermediate between the percept of each modality alone (McGurk & MacDonald, 1976). All sensory inputs are subject to both biases and variance, which means that the estimates from two different modalities are unlikely to be identical. Within the Bayesian framework, we can determine the most likely state that gave rise to these two sensory inputs. Specifically, the optimal combination of these two sensory inputs should be estimated by weighting each input by its reliability prior to averaging. In this manner, the more reliable modality contributes more to the final estimate. If we consider the estimation of the size of a held object, then we can combine the estimates of the size from both the visual and haptic sensory feedback. If we let the measurement from these two sensory channels be characterized by independent Gaussian noise, then the confidence that the object has size x relative to them is

$$P(v \cap h) = P(v)P(h) = \frac{e^{-\frac{(x-\bar{x}_v)^2}{2\sigma_v^2}} e^{-\frac{(x-\bar{x}_h)^2}{2\sigma_h^2}}}{\sigma_v\sqrt{2\pi}\ \sigma_h\sqrt{2\pi}} \tag{10.4}$$

The best estimate of the size of the object $\hat{x}$ is determined from the condition $0 \doteq dP(v \cap h)/dx$ as

$$\hat{x} = w \cdot x_h + (1-w) \cdot x_v, \quad w = \frac{\sigma_v^2}{\sigma_h^2 + \sigma_v^2} \tag{10.5}$$

where (x_h, σ_h) and (x_v, σ_v) are the mean and the standard deviation of the estimates arising from haptic and visual feedback, respectively, and w represents the relative weighting between visual and haptic information. In this manner, the information from each sensory modality is weighted by the relative variance of the estimate. For example, if both sensory

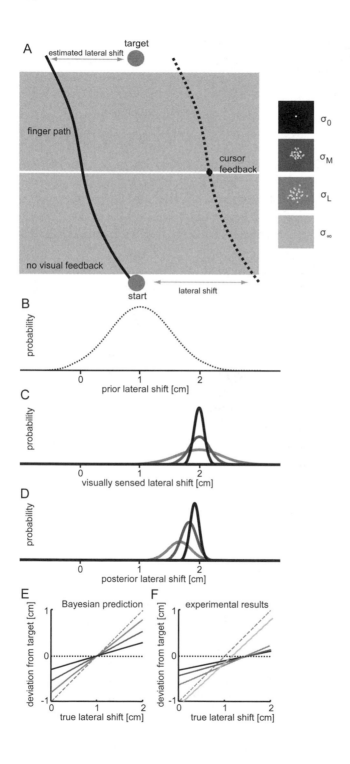

modalities have equal variance, the weighting factor will be one-half. On the other hand, if vision is removed (complete darkness), the variance associated with the visual estimate (σ_v^2) will become extremely large. This result leads to weighting that depends almost completely on the haptic estimate (as $w \rightarrow 1.0$). Similarly, removal of haptic feedback would produce a weighting factor equal to zero and lead to complete reliance on the visual estimate. This formula results in a variance of the optimal estimate ($\sigma_{combined}^2$) that is at worst as large as the smallest variance of the two sensory modalities:

$$\sigma_{combined}^2 = \frac{\sigma_h^2 \sigma_v^2}{\sigma_h^2 + \sigma_v^2} \tag{10.6}$$

Therefore, the sensorimotor system could obtain a more reliable estimation from combining the two estimates than from either sense alone. For example, the curves on the top of figure 10.3A show typical estimates of haptic and visual sensing (in this case, the estimates are unbiased but the underlying visual and haptic sizes are different). The haptic probability distribution is much wider than the visual probability distribution, as it is less accurate. Under such conditions, it is better to rely on vision. However, if the light level diminishes or the object is obscured, such as by smoke or fog, then haptic sensation will be more accurate than vision and should be relied upon more. Most of the time, when we are combining visual and haptic feedback about the same object, the sensations should match. However, there will always be biases in each modality, as they do not create estimates in the same sensory space. The question of whether the sensorimotor system uses this computational strategy to perform multisensory estimation was addressed in experimental studies with virtual objects (Ernst & Banks, 2002). Subjects were asked to estimate which of two paired rectangular objects was larger; both the visual and haptic size of the objects was simulated using a virtual reality workstation.

Figure 10.2
Bayesian integration of sensory feedback. (A) Subjects made arm movements with their finger moving from the start position to a target. Visual feedback of the location of the finger was provided only at the midpoint of the movement. The location was given by the actual position of the finger at that point in time plus a lateral shift, which varied randomly trial by trial. Subjects were instructed to place the cursor onto the target. The noise added to the visual feedback was increased by either providing a cloud of cursor positions or no visual feedback (blocks on right). (B) The trial-by-trial lateral shift added to the endpoint was Gaussian-distributed with a mean of 1 cm. (C) If, the actual lateral shift was 2 cm, subjects would perceive this shift as occurring at a location centered on 2 cm but with variability depending on the added visual noise. (D) If the perceived shift is combined optimally with the prior information (the distribution of possible shifts shown in B) then the required lateral shift would move closer to the 1 cm mean prior distribution as the noise increases. (E) Expected slope of deviations depending on the distribution of the visual noise if the sensorimotor system optimally combines the prior with the sensory information of location in a Bayesian manner. (F) The experimental results show the expected change in slopes with a higher reliance on the sensory information when the noise is small and a higher reliance on the prior when the noise is large (or no visual feedback is given). Adapted with permission from Körding and Wolpert (2004).

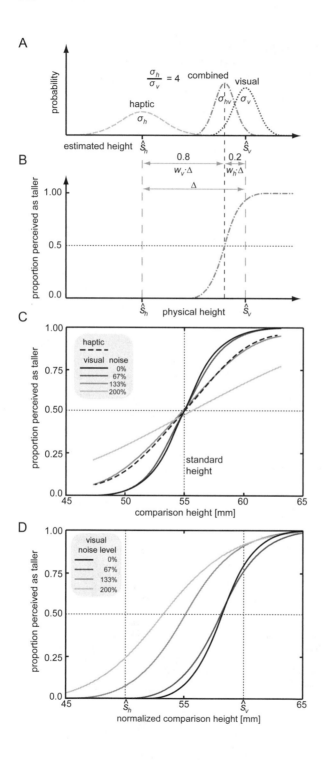

The study first determined the variability of each sensory modality independently. In haptic sensing, in which the relative size of grasped objects is estimated with eyes closed, differences in object size are difficult to discriminate when the sizes are similar, and therefore discrimination errors will occur more frequently as the sizes becomes more similar. By asking subjects to discriminate a wide range of object sizes and determining how frequently an object was perceived as larger than a reference object, a *psychometric function* can be generated that represents the probability that one object would be reported as larger than the other. Similarly, a psychometric function can also be generated from visual perception of object height. The slope of the psychometric function was found to be much steeper for visual sensing than haptic sensing (Ernst & Banks, 2002), indicating that visual sensing was more accurate (figure 10.3C). However, by adding noise to the visual feedback, the relative accuracy of the size estimates was reduced (greater variance reduces the slope of the psychometric function).

Ernst and Banks (2002) then examined how the sensorimotor system might combine the haptic and visual information to create a single estimate. In order to be able to examine the relative contributions of each modality, they created a virtual object with a larger visual than haptic size (figure 10.3B). If subjects were combining the two sensory modalities in an optimal manner, the judged height of the object should be closer to the visual than the haptic size. On the other hand, by adding noise to the visual feedback, subjects' judgment should shift toward the haptic size of the object.

Indeed, this was exactly what they found. For the zero noise condition, the slope of the psychometric function was steep with the point of subjective equality (where the function crosses 0.5) close to the visual size of the object. However, by adding noise to the visual feedback, the slope of the psychometric function was reduced and shifted toward the smaller haptic estimation (figure 10.3D). This observation implies that as the visual

---

Figure 10.3
Optimal combination of multisensory information. (A) The height of an object can be estimated by two different senses (e.g., haptic and visual). In this example, there are differences in the estimate (biases) from the two sensory modalities (the peak of the two estimates differs by Δ). The optimal combination of the estimates is produced by weighting (w) the relative contribution of each estimate by the relative variance of the estimate, with higher weighting for estimates with lower variance. In this case, the accuracy of the visual feedback is larger than for haptic perception (smaller width of Gaussian), so the optimal estimate weights the visual information more (0.8) than the haptic information (0.2). The combined estimate has lower variance than either estimate alone. (B) If an observer judged the height of this object based on the combined estimate, the psychometric function would be this sigmoid function (the integral of the Gaussian is a sigmoid function, the slope of which is defined by the width of the Gaussian). (C) Psychometric functions for haptic and visual discrimination, with each sensory modality examined independently. As noise was added to the visual feedback, the slope of the psychometric function decreased. (D) The size of an object was rendered larger in visual space than in haptic space through virtual reality. Four different levels of noise were added to the visual feedback. As the noise increased, the slope of the psychometric function decreased and the midpoint of the sigmoid moved towards the haptic estimate. This finding demonstrates that the sensorimotor control system adaptively weights multisensory information in an optimal manner to produce the estimate of the object size. Adapted with permission from Ernst and Banks (2002).

channel became noisier, a greater contribution to the sensory estimate was provided by the haptic modality, supporting the theory that the sensorimotor control system uses Bayesian statistics for multisensory integration (Ernst & Banks, 2002). This model of multisensory integration has also been demonstrated for estimation of location from visual and auditory cues (Körding et al., 2007) and estimation of body state from visual and proprioceptive information (Sober & Sabes, 2005; van Beers, Sittig & Denier van der Gon, 1996). In addition to integrating multiple modalities, a single modality can receive multiple cues about the same stimulus. Again, these different cues are combined optimally, such as when estimating object size from visual texture and motion cues (Jacobs 1999) and depth from texture and stereo cues (Knill & Saunders, 2003). This optimal combination of multiple sensory cues corresponds to Bayesian integration as described earlier. However, instead of sensory feedback being combined with prior information, the sensory information from two sensory modalities is combined to produce the optimal estimate.

10.2   Forward Models

In the previous examples, the sensorimotor estimates were formed from two sources of information about the state (e.g., the size of a cup) under static conditions. However, sensorimotor control acts continually in a dynamic and evolving environment. In particular, we need to maintain an estimate of the configuration of our body as we move so that we are able to generate the appropriate motor commands because the action produced by activating a muscle depends strongly on the current posture—for example, due to changes in muscle moment arms. Moreover, the dynamics—for example, limb inertia—also changes dramatically with posture. Errors in our estimate of the state of our limbs prior to movement can produce large movement errors (Vindras et al., 1998). Sensory feedback alone cannot provide accurate estimates of our current body state during movement, not only because the sensory signals are noisy but also because they represent past rather than current states as the result of delays. Therefore, if Bayesian integration is to be used to estimate time-varying states, it should take into account sensory delays as well as noise characteristics.

State estimation in a time-varying system can be carried out within the Bayesian framework. The likelihood assesses the probability of receiving the particular sensory feedback given the various states of the body. However, the prior now reflects the distribution of the most likely next limb states over all possible states given our best estimate of the current configuration. The physics of the world (and the kinematics of our body) means that the next state depends on both the current state and the motor command. The prior is calculated in a predictive manner using our previous state estimate and the motor command we have generated to change the state. A copy of the motor output (called an *efference copy*) and the current estimate of the state of the body can be used as inputs to a predictive or forward model that simulates the body dynamics (Wolpert & Miall, 1996; Wolpert,

Miall & Kawato, 1998). A forward model acts as a neural simulator of the dynamics of our neuromuscular system and the external world, predicting how our body moves in response to motor commands. However, forward model inaccuracy and noise contamination can cause the estimate of the future state to drift from the actual state over time unless sensory feedback of the actual state is continuously taken into account. Therefore, the forward model estimate must be combined with sensory feedback to produce an optimal estimate of the state of the body. The combination can be implemented in a Bayesian manner with the forward model prediction of the state acting as the prior. If this Bayesian integration is performed within a linear system, contaminated by (Gaussian white) noise in both the sensory input and motor output, it corresponds to a Kalman filter as introduced in chapter 9. The output of the forward model can be used for state estimation, prediction of sensory feedback, or predictive control, as described in the next sections.

Note the difference between a feedforward model learned by the sensorimotor control system to generate the forces corresponding to a desired action, as described in chapters 7 and 8, and the forward model presented in this section. Because the forward model can be seen as a mapping from the motor command to the predicted movement and the feedforward model as a mapping from the desired movement to the motor command, the feedforward model is sometimes called an inverse model.

State Estimation

Forward models are not only useful to counteract the effects of delays and noise but can also help in situations in which identical stimuli can give rise to different afferent signals, depending on the state of the system. For example, by modulating the γ-static and γ-dynamic drive to the muscle spindles, the sensorimotor system will receive different sensory signals in response to identical movements. To infer state in such situations, the sensorimotor system must take into account the motor output to interpret the sensory input. Only by combining multiple signals (e.g., γ-dynamic drive, γ-static drive, α-motoneuron drive) within a forward model of the system can inferences be made to interpret the state of the system when afferent signals are ambiguous. In fact, some evidence has suggested that muscle spindles may implement a forward model because they incorporate information related to movements up to 150 ms in the future (Dimitriou & Edin, 2010). Specifically, by recording the afferent firing in muscles during a sequence of finger taps, Dimitriou and Edin (2010) discovered that the afferent firing predicted the future velocity of the muscle-tendon complex. Such prediction could theoretically arise through an estimate of the motor commands and a model of the dynamics of the finger and hand. Alternatively, it might represent correlation between drive to α- and γ-motoneurons that would cause the spindle output created by γ-drive to appear advanced with respect to the later motion resulting from the α-activation. Whether this alternative explanation could account for an advance of 150 ms depends largely on whether the time required for excitation-contraction and transmission of force from the muscle to the skeleton through the tendon compliance matches the advance. In the case of rapid finger-tapping, compliance of the long tendons of the extrinsic finger muscles might introduce sufficient delay between change in muscle length and joint motion that the changes in muscle fiber length are significantly phase advanced with respect to joint motion.

Sensory Prediction

Prediction of future state can also be used to enhance perception (Wolpert & Flanagan, 2001). Specifically, if one can estimate the future state of the body based on the motor commands, then one can also estimate the sensory feedback that would result from the movement produced by those motor commands (figure 10.4A). Thus, the difference between the predicted sensory feedback (produced as a consequence of movement) and the actual sensory feedback can be calculated. Subtraction of the self-generated sensory feedback allows the sensorimotor system to accurately detect externally generated sensory feedback (figure 10.4B–D). Support for this theory was found in the vestibular nuclei for self-generated head movements in which the predicted head movement is subtracted from the afferent information (Roy & Cullen, 2001, 2004) such that the neural firing is insensitive to active head movement but remains highly sensitive to any passive movement.

Support for this was also found in investigating why one cannot tickle oneself. In (Blakemore, Frith & Wolpert, 1999), a robotic manipulandum was used to separate the self-generated action of tickling from the sensory effects of such motion. This separation was done by introducing either temporal delays or changes in the direction of movement between the self-generated motion and the subsequent robotic motion. In this way, the robotic interface introduces a discrepancy (either directional or temporal) between the action and the consequence of that action. It was found that when the robotic-generated tickle was produced in the same phase and direction as the generated movement, the sensation was reported to not be ticklish. However, as either the delay was increased or the direction of motion was rotated, the stimulus was reported as more ticklish. Thus, as the separation between the action and the consequence of that action increased (either spatially or temporally), the prediction by the sensorimotor control system of the sensory consequence of such action and the actual sensory feedback are no longer matched. This causes the sensorimotor control system to consider the sensory feedback as generated by external stimuli. Moreover, these results suggest that the prediction mechanism used for this sensory cancellation is both temporally and spatially precise. Similarly, sensory cancellation has been used to explain force escalation (Shergill et al., 2003). In this experiment, subjects were asked to take turns applying force to each other's fingers using a robotic manipulandum. Critically, the subjects did not know what rule the other subject used to determine the applied force. When both subjects were instructed to match the force that they felt, the force escalated, doubling in size from one trial to the next. This finding supports the idea that the subjects were not taking into account all of the self-generated force

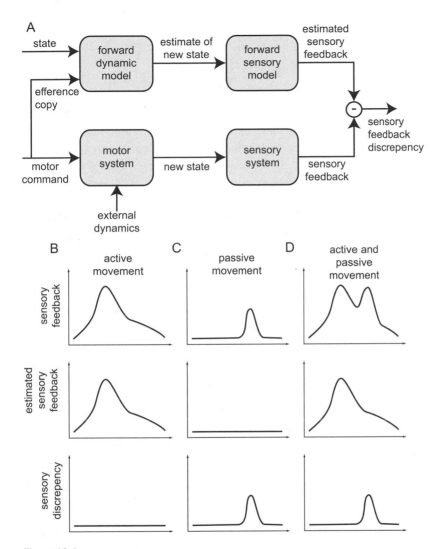

Figure 10.4
Forward models for sensory perception. (A) Block diagram of forward models being used to increase sensory perception of the external environment. The motor command is sent to both the motor system and a forward dynamic model (efference copy). The forward dynamic model uses both the copy of the motor command and the current state of the body to create an estimate of the next state of the body. This estimate is passed to a forward model of the sensory system, which simulates the expected sensory feedback from the body if the state changes as expected. The difference between the expected sensory feedback and the actual sensory feedback can be used to increase the sensitivity. Note that the forward dynamic and forward sensory models do not have to be separated. (B) In an active movement (with no disturbance), both the actual sensory feedback and the expected sensory feedback from the forward model produce similar responses, so the sensory discrepancy is close to zero. (C) When the limb is perturbed by the external environment (passive movement), the expected sensory feedback remains zero (as no motor command was produced), but the actual sensory feedback reflects the perturbation. The difference between the two accurately reflects the external perturbation. (D) The forward models allow subtraction of the predictable sensory feedback generated through motion produced by the motor command and that produced by external forces, which allows the system to remain highly sensitive to external perturbations by subtraction of any self-generated stimuli.

in their perception of the force that they were applying, resulting in large increases in the applied force.

Predictive Control

State estimation is particularly critical for control, as the delays in sensory feedback could otherwise result in inappropriate motor commands and large movement errors. For example, when visual feedback of the hand position is removed, the estimate of the hand position gradually drifts (Wann & Ibrahim, 1992). This drift can lead to large initial errors in the state estimation of the limb, which in turn can produce large errors in the endpoints of reaching movements (Vindras et al., 1998). It appears, from the predictable errors at the endpoints of the movements, that this error results from the highly nonlinear nature of the mechanical structure of the limb and muscles. Identical patterns of muscle activation for even small differences in the initial posture can lead to large differences in the final endpoint. These effects can be even more pronounced when the differences occur in the middle of movements, where the state changes quickly. Relying on delayed feedback for control would result in large errors in the state estimate and endpoint errors.

Conclusive evidence for forward models in the sensorimotor control system has been difficult to obtain, as the output of the forward model, the future state, is not measurable although it is used to control the system. As such, most studies have examined the effect of disrupting the forward model prediction on the motor output, by techniques such as increasing sensory feedback delays (Miall, Weir & Stein, 1993) or removing visual feedback (Mehta & Schaal 2002; Wolpert, Ghahramani & Jordan, 1995). Several lines of evidence suggest that the cerebellum is involved in state estimation for control (Miall et al., 2007). Disruption of the forward prediction was attempted by applying TMS over the cerebellum during a slow arm movement just prior to presentation of a visual target that subjects were required to intercept. When the cerebellum was stimulated, the movement was disturbed, causing errors in the ability to intercept the target. It appeared that during TMS trials, the interceptive reaching movements were planned using a hand position that was 140 ms out of date. This finding supports the theory that the cerebellum is involved in such predictive state estimation, without which the sensorimotor control system must rely on delayed feedback, resulting in errors.

Another line of evidence for forward prediction has investigated saccadic eye movements during reaching to probe-predicted hand position (Ariff et al., 2002; Nanayakkara & Shadmehr, 2003). Subjects were asked to visually track the position of their hand during reaching movements, while the eye gaze position was recorded. The gaze was found to be directed to a position advanced by an average of 196 ms compared to the hand position (Ariff et al., 2002). When arm position was disturbed by unexpected perturbations, they found that saccades were initially suppressed in the 100 ms period following the disturbance. Following this event, there appeared to be a recalculation of predicted hand position and the gaze again moved to a predictive position (150 ms in advance of hand position). In contrast, when the perturbation was accompanied by a change in the external dynamics, (i.e., a resistive or assistive force field), gaze position no longer reflected an accurate estimate of the future hand position, indicating that the prediction was impaired. This observation suggests that the prediction of future hand position was updated using both the sensory feedback of the perturbation and a model of the environment. When the model of the environment was incorrect, the sensorimotor system was unable to accurately predict hand position. On the other hand, after adaptation to the altered environment, saccades accurately predicted the actual hand position demonstrating that the model of the environment could be adaptively reconfigured (Nanayakkara & Shadmehr, 2003).

10.3 Purposeful Vision and Active Sensing

Although accuracy can be improved by combining sensory feedback with the prior belief or sensory information from other modalities, it can also be enhanced by optimizing resolution. This effect is particularly apparent with vision, in which resolution is highest at the fovea and drops dramatically as image location shifts away from the fovea. Because of this property, gaze must continually move from one location to another in order to optimize and maintain a high-resolution image of the world. For example, when we look at a person's face, we do not scan uniformly but shift our gaze between particular features (e.g., eyes, mouth, and nose) that are relevant to our purpose. Similarly, the eyes tend to gaze upon the target during reaching movements, as seen in figure 9.4A. The eyes normally saccade between gaze positions, during which the visual system receives no input; that is, visual perception is suppressed during saccadic eye movement. Observing the gaze direction during an action provides useful information on the features of interest and thus on the important aspects of the sensorimotor control during this action.

Visual information is used to explore the surrounding environment, providing selective information about features related to the tasks that we are performing. Changes in gaze position were examined during planned hand movements in order to understand how vision and action are coordinated in motor control. Subjects were asked to reach out and grasp a bar, move it to press a target switch while avoiding an intermediate obstacle, and return it to the original support surface (Johansson et al., 2001) (figure 10.5A,B). Analysis of the hand, object, and eye movements showed that subjects never fixated on the hand or the moving bar but instead fixated on landmarks critical to the control of the task. Specifically, subjects always fixated on the grasp site prior to grasping, the target, and the support surface where the bar was returned after performing the action (figure 10.5C,D). Obstacles along the movement path appeared to be optional landmarks, which attracted gaze only briefly if at all. Overall, shifts in gaze position always preceded hand movement, although these shifts were withheld until specific kinematic events had occurred. This finding suggested that vision was used by subjects for planning and monitoring critical kinematic

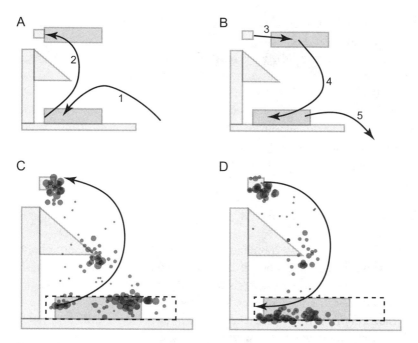

Figure 10.5
Active sensing. (A) Subjects were asked to move a small block (shaded gray block) from an initial start position (not shown) to a flat surface (1) and then around an object to touch a small target (2). (B) The second part of the task required the same movements in reverse: moving away from the target (3), avoiding the object and returning to the flat surface (4), and finally returning to the initial start position (5). (C) Position and frequency of eye fixation during the movements in (A). Size of circles represents duration of eye fixations for each location. (D) Position and frequency of eye fixation during the movements in (B). The results demonstrate that subjects focus on selected locations where the greatest accuracy is required, but rarely gaze at the hand or block itself during the movements. Figure based upon the results of Johansson et al. (2001) and used with permission.

events for verification of subgoal completion and demonstrates that vision is used as a major, active tool for action planning (Johansson et al., 2001).

Enhanced sensing is not confined to visual feedback. It has been shown that the afferent information arising from proprioception depends on the degree of γ-motoneuron activation, which in turn can vary with the task (Prochazka, 2011). Given that γ-motoneuron activity can enhance static or dynamic sensitivity of muscle spindles, this observation suggests that γ-drive could be used to selectively change the weighting of position-, velocity-, or acceleration-dependent signals. For example, if performing an action that requires high sensitivity (e.g., picking up a piece of soft tofu with chopsticks), the dynamic γ-drive could be increased, producing high sensitivity to any changes in velocity. This change would allow an impending change in the position of the fingers, which might indicate that the chopsticks were about to cut the tofu, to be countermanded. A critical feature of the sensorimotor system therefore may be to tune the sensors to the environment to produce sensory feedback that is optimally tuned to the task.

10.4   Adaptive Control of Feedback

As discussed in section 10.2 on forward models, the sensorimotor control system has access to an efference copy from the motor system for predictive modeling of future state. This prediction of the future state of the limbs during movement is useful for perception and control. Improvements in performance after the introduction of perturbations to limb movement suggest that the controller generating the corrective responses has access to the predictive state estimation. However, corrective responses can occur at a variety of delays, from rapid monosynaptic spinal reflexes to long-latency involuntary responses involving brainstem or cortical pathways and even longer-latency voluntary corrections. Although voluntary control is expected to have access to and use predictive state information, it is not obvious that involuntary feedback responses can incorporate such information. If rapid involuntary responses can incorporate predictive state information from a forward model, then not only should the feedback response depend on the state of the body, but because the forward model can also represent the dynamics of the environment, feedback responses could be modulated according to the task demands, as would be suggested by an optimal control model (Todorov & Jordan, 2002) as described in chapter 9. According to this model, motor control involves finding the best feedback control law for a given task that minimizes a cost function corresponding to the task (for example, maximizing accuracy and minimizing effort). The optimal control law therefore determines the set of feedback gains for the movement or task that achieves the minimal expected cost, which implies that feedback responses are not equally sensitive to all deviations. Optimal control predicts that the sensorimotor control system does not try to eliminate all variability but allows it to fluctuate freely in dimensions that do not interfere with the task (Todorov & Jordan, 2002) while minimizing it in the dimensions relevant for the task completion. In particular, the theory predicts that perturbations that do not interfere with the task completion will be ignored, whereas perturbations that interfere with the task will be appropriately coun-teracted (Scott, 2004; Todorov, 2004).

Task-Dependent Adaptation of Rapid Stretch-Dependent Responses

If the sensorimotor control system functions similarly to optimal feedback control, then rapid involuntary responses should have similar characteristics to voluntary commands, since both are produced by the same neural structures (Scott, 2004). Therefore, involuntary responses would be expected to exhibit the same task-dependent characteristics—adapting to the physical requirements of the task—as voluntary commands. Indeed, it has been shown that although the short-latency stretch reflex responses depend only on muscle stretch, the longer-latency stretch reflex responses change muscle activation in a manner that is similar to the later "voluntary" commands (Milner & Franklin, 2005; Pruszynski et al., 2009).

It has been shown that the long-latency stretch reflex responses are organized in a manner that takes into account the acceleration-dependent interaction dynamics so as to appropriately counteract the applied torques rather than the imposed motion (Kurtzer, Pruszynski & Scott, 2008, 2009; Lacquaniti & Soechting, 1986a, 1986b; Soechting & Lacquaniti, 1988). In the recent studies by Kurtzer and colleagues, torque perturbations were applied to the shoulder and elbow joints such that either only the shoulder or only the elbow was moved. Applying a perturbation torque at either joint alone produces a torque about the other joint due to the interaction forces between the two segments of the arm causing both limb segments to move (figure 10.6A). By selecting an appropriate combination of shoulder and elbow perturbation torques, the motion can be restricted to one of the joints, such as the elbow (figure 10.6B–D). However, if the neuromuscular system responds by generating a compensatory torque at the moving joint only, then the other joint will move due to the interaction forces. Appropriate compensation for the overall disturbance must take into account the interaction torques and not only the externally imposed motion. Although the short-latency monosynaptic stretch reflexes indeed reflect the muscle stretch produced by the externally imposed motion, the longer-latency responses counteract the interaction torque as well (figure 10.6E). Moreover, such effects were found to scale with the size of the imposed torques (Crevecoeur, Kurtzer & Scott, 2012).

The authors interpreted the results as indicating that the sensorimotor system must use information from both joints to resolve the local ambiguity in the stimuli in order to produce an appropriate response (Kurtzer, Pruszynski & Scott, 2008). That is, measurement of either the elbow or shoulder motion alone cannot resolve the underlying applied torques. The results were interpreted as indicating that these sensory signals are integrated with a model of the limb dynamics to produce the appropriate response at the shoulder. However appealing this view might be, it is not the only explanation. The biarticular muscles that span the elbow and shoulder joints could be used to resolve the ambiguity considered to be present at the level of each joint. Specifically, if the elbow joint undergoes motion while the shoulder does not, the biarticular muscles will necessarily change length. This change in muscle length creates a change in the force at both joints that will be sensed by the Golgi tendon organs in the muscles. Such signals are transmitted by large-diameter afferents with conduction speeds close to those of muscle spindles. Thus, such signals from muscle afferents acting at the shoulder already provide information about the direction and magnitude of the motion at the elbow. Thus, there are a variety of combinations of the sensory signals that could disambiguate the underlying torque perturbations creating the joint motion. Such processing could potentially occur even within the spinal circuitry. Whether such spinal circuitry could be considered as an internal model of the limb

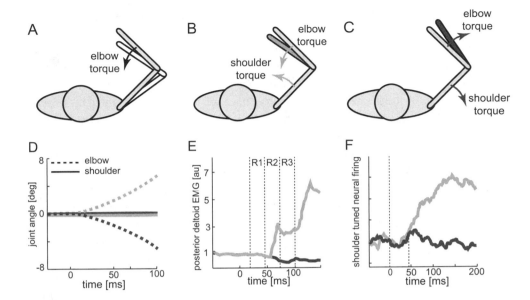

Figure 10.6

Feedback responses incorporate knowledge of limb dynamics. (A) If torque is applied at the elbow only, movement will occur at the elbow directly and at the shoulder through segment interaction forces. (B) By applying a specific combination of torque at the shoulder and elbow the motion imposed through segmental dynamics at the shoulder is prevented leading to only flexion of the elbow. Note however, that if the muscles resisted only the imposed elbow torque, flexor motion at the shoulder would occur, as there would be a net shoulder flexor torque. (C) Similarly, a specific combination of extension torques applied to the elbow and shoulder can lead to extension of the elbow only. (D) Joint kinematics after the onset of the external joint torques for the conditions shown in (B) and (C). The imposed torques produce significant rotation of the elbow but no motion at the shoulder. (E) Changes in activation of the posterior deltoid muscle (shoulder extensor). As the muscle is not stretched by the perturbation, no short-latency responses are seen (R1). However, divergence in the responses for the two types of perturbations occurs in the middle of the first long-latency period (R2) and is maintained throughout R3 and during later voluntary muscle activation. This response compensates for the initial (acceleration-dependent) shoulder torque imposed by the perturbation in each condition. (F) Neural firing rates in primary motor cortex for shoulder tuned neurons. Shoulder tuned neurons modulate their activation only when resisting a steady-state imposed shoulder torque. The task-dependent changes in neural firing for the two perturbations occurs about 10 ms prior to the changes in muscle activation, suggesting the direct involvement of primary motor cortex neurons in these task-dependent feedback responses. Figure based on Kurtzer, Pruszynski, and Scott (2008) and Pruszynski et al. (2011) and adapted with permission.

dynamics depends on the definition or representation of an internal model that one is willing to accept.

In order to examine the neural circuits that underlie such responses, Pruszynski et al. (2011) used the same experimental design while recording the neural firing of primary motor cortex neurons. They constrained their analysis to neurons that fired selectively in response to steady-state shoulder torque in order to examine neurons thought to be involved primarily in voluntary motor control of the shoulder joint. Joint torque perturbations were then applied to the shoulder and elbow such that the motion only occurred at the elbow joint. Neural firing in the shoulder-tuned neurons increased within 20 ms of the onset of the perturbing torques (a response that was not specific to the direction of elbow motion) and began to exhibit perturbation specific tuning to the underlying perturbations within 50 ms of the perturbation onset (figure 10.6F), that is, producing an appropriate increase or decrease in neural firing depending on the direction of the shoulder torque produced by the elbow motion despite the absence of motion at the shoulder joint (Pruszynski et al., 2011). This result suggests that these motor cortex neurons (involved in voluntary control) receive and process sensory information from multiple muscles crossing the surrounding joints allowing them to produce appropriate responses to the motion. Although the timing of these perturbation specific responses were early enough to contribute to the task-dependent long-latency reflex responses (Kurtzer, Pruszynski & Scott, 2008) and mirrored the muscular responses, these results do not demonstrate direct causality in the long-latency responses. In order to examine this concept further, the authors applied TMS over the motor cortex to excite the local cortical circuitry (Pruszynski et al., 2011) during the joint torque perturbations. Single-pulse TMS excites cortical circuitry, causing a supralinear response when synchronized with the long-latency stretch response but only a linear response when synchronized with the short-latency reflex that involves purely spinal circuitry (Day et al., 1991). Pruszynski et al. (2011) combined this technique with perturbations that produced elbow but not shoulder motion, finding supralinear responses in the shoulder muscle activity only when the TMS was timed to interact with the long-latency feedback responses. The results suggest that the cortical circuitry is involved in producing the task-dependent feedback response to the perturbations. This interpretation has been supported by several other studies using combined TMS and limb perturbations in human subjects, which have also implicated the involvement of cortical neurons in the task-dependent increase in feedback gains (Kimura, Haggard & Gomi, 2006; Shemmell, An & Perreault, 2009). However, it is not possible to completely rule out the possibility that these responses are mediated by TMS-induced changes in excitability of brain stem or spinal reflex circuitry through descending inputs that do not affect the monosynaptic circuit. Nevertheless, these findings, when combined with prior studies implicating the transcortical pathway in the long-latency feedback response (Matthews, 1991) and task-dependent responses in the cortical neurons prior to the long-latency feedback response

(Pruszynski et al., 2011), support the involvement of cortical circuitry in these task-dependent responses.

Not only do the responses of these long-latency stretch reflexes reflect compensation for the interaction dynamics of the multijoint limb, but they have also been shown to be tuned to task differences. When subjects make separate reaching movements simultaneously with their two arms (figure 10.7A), a perturbation applied to one hand elicits a reflex response in the perturbed arm only that acts to return the hand back to the original

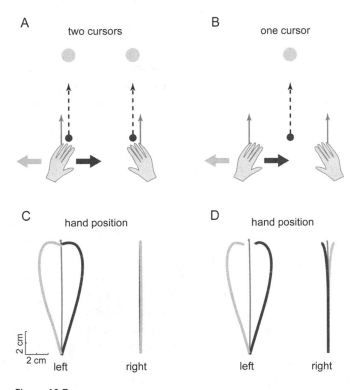

Figure 10.7
Feedback responses incorporate sensory information from both limbs for optimal responses to perturbations. (A) Subjects make reaching movements with their two hands, each of which independently controls a cursor (black dots). During the movements, the left arm could be perturbed either to the left or right by applying force to the hand (arrows). (B) Subjects make reaching movements with their two hands, which jointly control a single cursor, that represents the average position of the two hands (black dot). During the movements, the left arm could be perturbed either to the left or right by applying force to the hand (arrows). (C) The trajectories of the two hands during the movements in (A). The applied force perturbs the left hand, which then produces appropriate compensation at the end of the movement to reach the target. No responses are seen in the right arm. (D) The trajectories of the two hands during the movements in (B). The applied force perturbs the left hand, which then produces appropriate compensation at the end of the movement. The compensation is smaller than under the two-cursor condition, as the right arm also produces a corrective response such that the joint cursor reaches the target. Figure adapted with permission from Diedrichsen (2007).

trajectory (figure 10.7C). However, if the two arms together control a single cursor located at the spatial average of the two hands (figure 10.7B), the same perturbation applied to one hand elicits feedback responses in both arms to appropriately adjust the cursor's position (figure 10.7D) (Diedrichsen, 2007). These results might be interpreted as arising from an optimal controller that divides the required change in the control signal between the two limbs. Regardless of the interpretation, these results demonstrate that the long-latency stretch reflexes are adapted to complexities of the task and do not function simply to reduce the locally induced errors. Further demonstrations of the adaptive responses of long-latency stretch reflexes to perturbations of both arms have been shown during object manipulation (Dimitriou, Franklin & Wolpert, 2012). Subjects were asked to control the orientation of a virtual rigid bar using both arms. A variety of perturbations were then applied to both limbs. The magnitude of the response to these perturbations demonstrated that the long-latency stretch reflex incorporated information from both limbs to counteract the overall disturbance. This finding suggests that these rapid feedback responses combine sensory information from multiple joints along with a model of the task. Together, these and many other studies (see Pruszynski & Scott [2012] for a recent review) suggest that the feedback system has access to predictive models of the environment, allowing rapid control that takes into account the dynamics of the limbs.

Task-Dependent Adaptation of Rapid Visuomotor Responses

Although stretch-dependent muscular feedback may be the most extensively investigated feedback mechanism that produces corrective responses to perturbations, visual feedback can also elicit muscular responses at short latencies. Rapid motor responses to visual feedback have been found in responses to shifts in the target location (Prablanc & Martin, 1992; Soechting & Lacquaniti, 1983), shifts in the visual hand position (Sarlegna et al., 2003; Saunders & Knill, 2003, 2004) and shifts in the visual background (Gomi, Abekawa & Nishida, 2006; Saijo et al., 2005). These stimuli produce responses in the limb musculature starting approximately 100 ms after the visual shift and have been shown to be involuntary in nature (Day & Lyon, 2000; Franklin & Wolpert, 2008). Such rapid visuomotor responses require an estimate of the state of the limb to activate the muscles to respond appropriately to the perturbations because visual feedback contains no muscle-specific sensory information. Instead, the sensorimotor control system must combine the visual information about the desired corrective response in external space with the state information about the current position and motion of the arm to estimate the required muscle activation.

Similar to stretch reflexes, visual feedback responses exhibit task-dependent modulation for both perturbations of the target and hand locations. Liu and Todorov (2007) demonstrated that the response to perturbations of the target during reaching movements varies depending on the time in the movement at which the target jump occurs, with the gain decreasing toward the end of the movement. In another study the visual environment was manipulated in such a way that a sensory discrepancy between the visual location and the proprioceptive location of the hand was introduced in the middle of the movement. This sensory discrepancy could be either task relevant or irrelevant depending on whether subjects were required to compensate for the discrepancy. The visuomotor reflex gain was probed throughout the movement using perturbations of the visual location of the hand. These probes indicated that the reflex gain was increased in task-relevant conditions but decreased in task-irrelevant conditions (Franklin & Wolpert, 2008). It has also been shown that target shape modulates the size of the visuomotor reflex response such that larger responses to perturbations are produced in the directions with the smallest tolerance for errors (Knill, Bondada & Chhabra, 2011). More recently, it has been demonstrated that these visuomotor reflexes are excited during the initial learning of novel dynamical force fields, suggesting that these feedback gains are upregulated with increased uncertainty in the knowledge of the external dynamics to compensate for disturbances and ensure skillful motion (Franklin, Wolpert & Franklin, 2012).

Overall, these studies show that feedback mechanisms produce responses to perturbations demonstrating many of the hallmarks of voluntary motor control and suggest that the same brain structures may be involved in both control systems. Feedback responses appear to incorporate limb state information and are modulated according to the dynamics and task requirements. This finding suggests that feedback mechanisms can incorporate, or have access to, predictive models of the limb state and internal models of the task dynamics.

10.5 Summary

Sensing is not a stereotyped process but adapts to the neuromuscular system, the environment, and the task demands as appropriate. The brain combines noisy sensory information from multiple modalities in a statistically optimal way, weighting the information by its reliability in order to produce an estimate with the minimum variability. The sensorimotor control system combines the efferent copy of the motor command with a model of the body and environment to produce a predictive estimate of the current and future state of the body in order to avoid having to rely exclusively on delayed feedback. By combining this estimate with delayed sensory feedback, the predictive estimate is never far from the actual state of the body. Adaptive sensing indicates that the sensorimotor control system produces motor outputs designed to increase the sensitivity of afferent feedback, producing high-resolution information about only a few key elements relating to the task. Sensory feedback is adaptive to the environment, demonstrating adaptive responses to perturbations that vary depending on the task relevance of the information.

11 Applications in Neurorehabilitation and Robotics

In previous chapters, we have described what is known about human neuromechanics and its implications for the control of human motion. This knowledge can be applied in a wide variety of settings. For example, in chapter 6 we described how this knowledge has been used to animate human figures or humanoid robots in a way that appears natural. In this chapter, we illustrate two other important areas of application, namely *neurorehabilitation* and *robotics*. On one hand, knowledge of the neurophysiology reviewed in this book can help us interpret motor behaviors in the context of neurorehabilitation and predict the dynamics of the interaction between a human and a robot. On the other hand, the computational models presented in previous chapters may be used to design efficient rehabilitation protocols and robotic controllers, as well as to provide efficient human-machine interaction algorithms. This chapter presents a series of case studies in neurorehabilitation and robotics related to the material of this book, rather then a comprehensive overview of these fields.

11.1 Neurorehabilitation

The proportion of individuals with motor impairments due to neurological disorders (e.g., stroke, Parkinson's disease, cerebral palsy, spinal cord injury) is increasing worldwide, particularly because of aging populations. Stroke is one of the leading causes of adult disability, affecting approximately one in five hundred every year. Although stroke can affect anyone at any time, the risk increases with age and doubles each decade after age fifty-five. In developed countries with aging populations, this has become a major health issue (Hollander et al., 2003). Stroke is a sudden disabling attack or loss of consciousness caused by an interruption in the flow of blood to the brain (ischemic stroke) or by the rupture of blood vessels in the brain (hemorrhagic stroke). As a result, the affected area of the brain is unable to function, often leading to inability to move limbs on the side of the body opposite to the affected area, inability to understand or formulate speech, or an inability to perceive half of the visual field.

In the acute phase after a stroke, the ambulatory treatment is focused on stabilizing the vital functions. In the following six to twelve weeks, patients are provided neurorehabilitation at the hospital in the form of one-on-one therapy with physiotherapists (figure 11.1A), which facilitates spontaneous recovery. Due to limited human and financial resources, patients are usually sent home when they are able to walk, even though their upper limb function is often severely impaired, affecting quality of life and independence in activities of daily living (ADL). However, there is evidence that patients would be able to improve upper limb motor function with more therapy (Heller et al., 1987; Wade et al., 1983). Injured sensorimotor circuits in the brain can reorganize with extensive training, as demonstrated by imaging studies (Takahashi et al., 2008; Bosnell et al., 2011). This neural plasticity may have similarities to motor learning in healthy subjects as described in chapters 7 and 8.

11.2   Motor Learning Principles in Rehabilitation

At the end of the last century, there were two principal approaches to neurorehabilitation (Winstein & Wolf, 2008): (1) training of typical ADL and (2) neuromuscular training of specific movements. However, training of ADL or specific movements per se may not benefit the patient much if the training is not appropriately adapted to the patient's motor impairments. This issue is considered in *task-oriented training* (ToT), which is an integrated view of neurorehabilitation that has recently emerged as the dominant approach to motor restoration. In ToT, rehabilitation is focused on the acquisition of skills for performance of meaningful and relevant tasks for each patient (Winstein & Wolf, 2008). Behaviors that can be beneficial for improving the motor function of the patient are selected and adapted for training. In general, ToT has been shown to be of more benefit to patients with moderate impairment than those with severe impairment. Therefore, care should be taken to ensure that the complexity of the training tasks does not exceed the patient's capabilities.

As motor learning in healthy subjects and neurorehabilitation of sensorimotor impairments after stroke are both believed to be based on neural plasticity, Kleim and Jones (2008) proposed a set of principles for efficient task-oriented neurorehabilitation based on systematic studies of neural plasticity in healthy motor learning such as those discussed in chapters 7 and 8. These principles can be summarized as follows (Winstein & Wolf, 2008):

• *Use it or lose it:*   Improvement of the motor condition comes from training the limb.

• *Use it and improve it:*   Patients should practice what they can already do and adapt the intervention as motor function evolves. Training should be challenging and optimally adapted to the patient's current motor condition.

• *Specificity:*   It is important to train specific motor skills, as these are represented in memory in a highly specialized way.

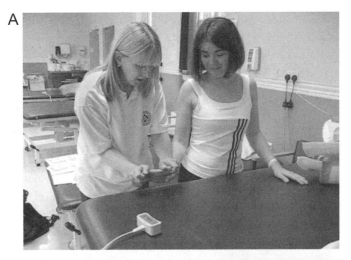

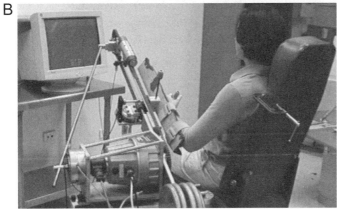

Figure 11.1
Post-stroke neurorehabilitation. (A) Traditional therapy consists of one-on-one exercises performed with a phys-iotherapist. (B) A solution to the increasing number of stroke survivors and limits to the number of physiothera-pists and financial resources may consist of using dedicated rehabilitation robots for repetitive training and on-line assessment of motor function, adapting assistance according to the patient's state of motor recovery. Adapted with permission from Reinkensmeyer, Schmit, and Rymer (1999).

• *Repetition matters:*   Learning requires sufficient repetition.

• *Intensity of the therapy matters:*   Mere repetition of tasks that are too easy for a patient will not induce learning. The dose, frequency, and duration of training are important parameters of learning efficacy. For example, as mentioned in chapter 7, randomization of the tasks improves learning.

• *Timing of the intervention matters:*   In particular, recent clinical evidence has shown that post-stroke rehabilitation is most effective in the *subacute phase*, as soon as possible after the patient's vital functions have been stabilized.

• *Salience matters:*   As the use of the limbs for everyday tasks is what patients would most like to recover, ToT should include a representative set of such tasks.

• *Transference:*   Training can generalize to similar tasks. Generalization may be facilitated by slightly varying the conditions during training. Note that this principle is not contradictory to specificity but complementary to it.

• *Interference:*   Learning one skill can interfere with the acquisition of other skills. Therefore, ToT should take place early enough to prevent compensatory strategies from developing, which may be difficult to replace with "correct" behaviors, and it should focus on training skills that lead to improvement of motor function.

An as-yet unresolved issue, which may be addressed through motor learning studies similar those presented in chapter 7, is the identification of an optimal set of motor training skills that are sufficiently specific, minimize interference, and together generalize to a variety of important ADL.

11.3   Robot-Assisted Rehabilitation of the Upper Extremities

Dedicated robots that can interact with human motion, physically assist movement, accurately measure performance, and precisely control training are being used to carry out neurorehabilitation of post-stroke patients (figure 11.1B). Combining robotics with virtual reality workstations can provide safe and motivating therapeutic games to engage patients in therapy. Rehabilitation robots can provide continual assessment of changes in motor function (Balasubramanian et al., 2012) through performance measures or through measurement of the response to perturbations, such as those used to assess spasticity (Palazzolo et al., 2007). These capabilities are needed to develop an evidence-based quantitative approach to neurorehabilitation that can lead to optimal training techniques for speed and extent of recovery (Balasubramanian, Klein & Burdet, 2010). Because rehabilitation robots can provide intense goal-specific training, permit even weak patients to train, and acquire quantitative information about the patient's motor condition, they are ideally suited for implementing ToT.

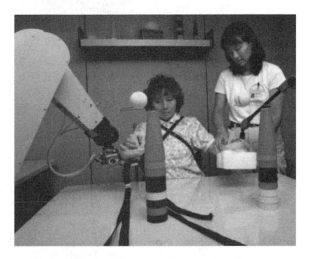

Figure 11.2
A stroke survivor training with unstructured arm movements in 3D space using mirror-image therapy, in which the movement of the healthy (left) arm is recorded (by a mechanical measurement system) and used to control the affected (right) arm with the robot. Reproduced from Burgar et al. (2000) with permission.

Various robotic systems have been developed over the past fifteen years for neurorehabilitation applications. Some of the earliest systems that have been used in rehabilitation of arm movements are the MIT Manus (figure 6.4) and the MIME system (figure 11.2) at Stanford (Prange et al., 2006; Kwakkel, Kollen & Krebs, 2008). These robots have been used to implement a number of different upper limb training modalities (figure 11.3), including:

• *Passive training,* in which the limb is driven by the robot (figure 11.3A). This training provides the patient with proprioceptive sensory feedback in the absence of voluntary muscle activation and can be used to stretch muscles and increase the passive range of motion.

• *Path guidance,* which prevents the patient from making hand path errors (figure 11.3B). This modality provides the patient with proprioceptive sensory feedback of errors in force direction but does not correct muscle activation patterns.

• *Active control* with assistive force (figure 11.3C) to allow patients to increase speed or complete difficult movements or with resistive force (figure 11.3D) to help increase muscle strength. This modality provides more normal proprioceptive feedback during movement.

• *Error augmentation* with an unstable force field (figure 11.3E) producing amplified sensory (proprioceptive) feedback of errors, forcing the patient to overcorrect faulty muscle activation patterns.

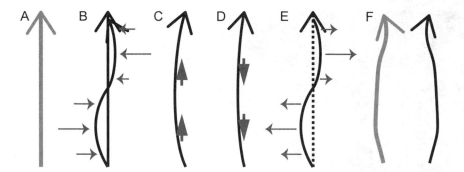

Figure 11.3
Training modalities used in studies of robot-assisted rehabilitation. (A) passive, (B) path-guidance, (C,D) active control (C: assistive, D: resistive), (E) error augmentation, (F) mirror-image modality.

• *Mirror-image modality,* in which voluntary movement of the unimpaired limb "teaches" the impaired limb (figure 11.3F) and provides a means to compare sensory feedback from active (unimpaired) and passive (impaired) limbs. This modality is also illustrated in figure 11.2.

The results of clinical trials on robot-assisted arm training can be summarized as follows (Prange et al., 2006; Kwakkel, Kollen & Krebs, 2008):

• Robot-assisted therapy is at least as effective as conventional therapy performed at the same intensity.

• The clinical improvements following intensive robot-assisted therapy of chronic patients are statistically significant but possibly not large enough to be functionally meaningful.

• Passive training is insufficient; active participation is required to obtain significant functional improvement.

• Training planar arm movements does not transfer well to functional tasks, such as manipulation of objects with the impaired limb.

• As the hand is required for ADL such as eating, manipulating objects, or handwriting, more recent approaches have focused on training of hand function (Balasubramanian et al., 2010), either in addition to training the arm or without arm movement. Initial findings are similar to those listed previously for arm movement training, but improvement has been found both in hand and arm function even when only the hand was trained (Lambercy et al., 2011).

11.4 Application of Neuroscience to Robot-Assisted Rehabilitation

Although neurorehabilitation is different from motor learning in healthy subjects, studies of human motor learning can contribute to the development of efficient post-stroke rehabilitation strategies in various ways. For example:

• Experiments with healthy subjects learning a novel task can provide insights that may be useful for rehabilitation. Neurorehabilitation and motor learning are likely achieved through similar mechanisms of neural plasticity, so strategies that do not work well in normal motor learning are unlikely to work better in neurorehabilitation. This approach is ethically satisfying as such testing enables us to rule out inefficient strategies without comprising the treatment of post-stroke patients. On the other hand, strategies that increase the rate of motor learning or lead to improved performance may serve as templates for the development of more effective protocols for neurorehabilitation.

• Neuroscience research can be designed to investigate neural structures and processes involved in rehabilitation. In particular, techniques such as functional brain imaging and TMS can be used to reveal the mechanisms of functional recovery and determine which regions of the brain are necessary to achieve a specific level of recovery. This approach may eventually lead to the development of evidence-based, subject-specific rehabilitation protocols.

11.5 Error Augmentation Strategies

A first example of a strategy inspired by motor learning studies is illustrated by the work of Patton, Kovic, and Mussa-Ivaldi (2006), who developed a robot-assisted protocol for post-stroke neurorehabilitation based on learning the type of novel dynamics discussed in chapter 7. In their initial study, they had post-stroke subjects train in either a clockwise or counterclockwise velocity-dependent (curl) force field. In this study, they showed that in directions where movements began with significant errors relative to the target direction, improvements occurred only when the training forces magnified the initial errors. Subsequently, they created a subject specific force field designed to lead to formation of an internal model with aftereffects that would correct for the motor control deficits (Patton, Kovic & Mussa-Ivaldi, 2006). When an impaired movement deviates from a straight-line path to the target, as shown in figure 11.4A, a machine-learning algorithm is used by the robot to determine the force that it must apply at the hand to restore the straight-line path. The opposite force is then imposed on the hand during a training session in which the patient performs the impaired movement (figure 11.4B). Following training, the robot stops applying force to the hand and aftereffects are observed that show an adaptive effect that restores the straight-line path (figure 11.4C). A critical question is whether the learned behavior persists. For instance, aftereffects in healthy subjects are washed out within a few trials of returning to the original dynamics. On the other hand, learning protocols based on knowledge of how motor memory works as described in chapter 7 may be used to improve retention. Thus, if a newly learned motor behavior is beneficial to a post-stroke subject it might be consolidated in motor memory with appropriate training techniques. However, identification of optimal motor behavior for a post-stroke subject is not straight-forward given that optimal motor behavior for his or her condition may be different from that of a healthy subject.

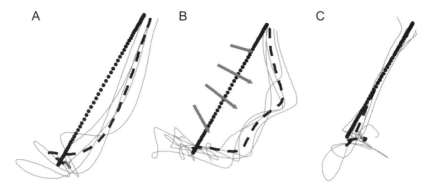

Figure 11.4
Correcting movements of post-stroke patients through learning with error-enhancing force fields (Patton &
Huang, 2012). (A) Initial movements of a post-stroke patient. (B) Movements under error enhancement (arrows
indicate force applied by the robot) and (C) path straightening seen in after-effect trials following training with
error enhancement.

As rehabilitation is performed according to the sensory signals provided to the patient's
CNS, a possibility for speeding up sensorimotor recovery and augmenting motor capacities
consists of manipulating sensory feedback (Patton & Huang, 2012). In a study on walking
(Emken & Reinkensmeyer, 2005), subjects learning how to counteract a force disturbance
increased their rate of learning by 26 percent when a disturbance was transiently amplified.
Similarly, visual feedback of movement error can be amplified in order to promote learning. Wei et al. (2005) had neurologically normal subjects learn a visuomotor rotation while
the visual feedback of the hand was manipulated to amplify its lateral displacement from
a straight line joining the start and target positions. An amplification factor of 2 led to
faster learning than an amplification factor of 1 or an amplification factor close to 3.

Although there is some evidence that these augmentation strategies can improve learning, they use a reference trajectory specified by the experimenter relative to which error
is computed. We would like to develop methods such that patients can train by exploration
rather than relying on feedback of performance. Huang and Patton (2012) created a
mechanical environment for amplifying error and promoting exploration of the sensorimotor space by having subjects move a robot programmed to generate negative viscosity,
similar to that introduced by Milner (Milner & Cloutier, 1993; Milner, 2004). The idea is
that the negative viscosity increases the range of movement velocities and the amplitude
of errors experienced during training by assisting and perturbing movement. Post-stroke
subjects were instructed to move the arm within a prescribed workspace while holding the
robot, after which they performed circular evaluation trials in the null field. Subjects who
trained with the negative viscosity showed a significant improvement in their ability to
move along the circular target path during evaluation trials compared to control subjects
who trained only in the null field (Huang & Patton, 2012).

Although more work will be needed to determine the most appropriate modifications to sensory inputs for promoting motor learning, these results demonstrate promising approaches for improving motor recovery following stroke. There are several possible explanations for the beneficial effects of sensory augmentation. First, there is strong evidence that different brain areas are involved in processing large and small errors during motor learning. In particular, cerebellar damage impairs the ability to learn from large errors (Criscimagna-Hemminger, Bastian & Shadmehr, 2010). Second, as some learning and motor recovery is based on error minimization, augmenting feedback error may provide larger learning gains. Third, augmenting error may provide positive motivation to post-stroke patients by creating the illusion that their recovery is progressing faster than in reality, which might allow patients to gradually increase their range of movement without being too disappointed by poor initial performance. Finally, error augmentation may enable an impaired sensory system to perceive otherwise imperceptible errors by supplementing the feedback provided by one sensory modality with that provided by another. For example, Lambercy et al. (2011) used visual feedback augmentation to train forearm pronation/supination, supplementing proprioception. However, some caution must be exercised in promoting error amplification as a training tool because it has been shown that learning with large errors leads to poorer retention (Huang & Shadmehr, 2009; Klassen, Tong & Flanagan, 2005) and less robust generalization (Kluzik et al., 2008) than training with small errors.

11.6 Learning with Visual Substitution of Proprioceptive Error

We have seen in chapter 7 that force field learning relies on kinematic error, which can be detected by various sensory modalities, including proprioception and vision. For the purpose of designing simple rehabilitation tools, it is worth examining the effectiveness of providing only visual feedback in motor learning. For example, as reaching movements roughly follow a straight-line path, one strategy in designing rehabilitation robots for reaching has been to create a simple linear guide (figure 11.1B; Reinkensmeyer, Schmit & Rymer, 1999; Lam et al., 2008; Yeong et al., 2009). Such robots are mechanically simpler and more compact than multiple DOF robots. However, the absence of lateral error imposed by the mechanical constraints could prevent efficient learning.

Therefore, we investigated whether the lack of proprioceptive error feedback could be compensated by providing visual feedback of the movement that would have occurred had the arm not been constrained (Melendez-Calderon et al., 2011). The force exerted by the subject was recorded during movement in a constraining haptic channel and used to compute the trajectory that would have resulted from this force if the hand had been moving in a VF, using an inverse dynamics model of the arm and the force field. The subject was given online visual feedback of that trajectory. Note that this situation represents a mismatch between visual and proprioceptive feedback; proprioception detects zero

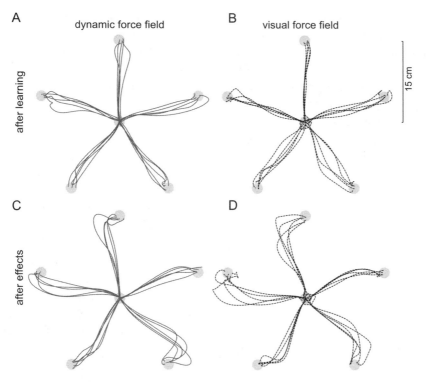

Figure 11.5
Movements of typical subjects who have learned to move in a curl force field either by training in the force field (A) or by visual feedback of a virtual representation of the force field when movement was constrained laterally by a haptic channel (B). The corresponding aftereffects assessed in a null field are shown in (C) and (D), respectively.

lateral error as the movement is constrained by the channel, whereas the subject sees lateral movement according to the force exerted on the channel.

The movement paths after training are shown in figure 11.5. Remarkably, subjects who were trained in a haptic channel with only visual feedback moved in a VF with trajectories that were similar to those of subjects who were trained in the actual force field (Melendez-Calderon et al., 2011). Furthermore, after-effect trajectories were also similar, suggesting that the two groups of subjects had learned the same dynamics. This finding suggests that a similar feedforward model of the task was learned with the virtual force field as with the real one. Although learning is slower with only visual feedback and the learning processes in the two cases are likely different (Melendez-Calderon et al., 2011), performance after learning was essentially the same. In addition, training in the haptic channel with visual feedback of the simulated force field aided retention of learning. When subjects initially learned by training in the true force field and subsequently continued their training in the haptic channel, the force that they applied to the haptic channel dropped within a few trials if the visual feedback represented the true hand motion. However, the applied force did not significantly change if the visual feedback represented motion in the simulated force field (Melendez-Calderon, 2011).

Note that this method can be generalized to other types of constraints. In particular, many studies of robot-assisted arm rehabilitation have been carried out using a robotic interface constrained to move in the horizontal plane. Although such interfaces provide weight support that can help post-stroke patients increase their range of motion (Ellis, Sukal-Moulton & Dewald, 2009), removing such support may create aftereffects. As the patients do not learn to support the weight of their arm, they may, for example, initially spill their glass of water. By including visual feedback of a virtual gravitational field during rehabilitation using the principles described previously, it may be possible to avoid this effect and allow the patients to gradually learn to support the weight of their arm.

11.7 Model of Motor Recovery after Stroke

In chapters 7 and 8, computational models of motor learning in healthy humans were developed. Similar models of motor recovery after stroke could be developed to better understand the nature of the recovery process and to plan and deliver therapy. Optimal therapeutic protocols may be developed based on subject-specific computational models of recovery.

In an effort to understand the dynamics of motor recovery and the role of physical assistance provided by a robot during therapy, Casadio and Sanguineti (2012) developed and tested an iterative linear dynamical model of motor recovery after stroke based on the model of force field learning in healthy subjects presented in section 7.2. In this state space model of robot-assisted reaching:

$$y^k = z^k + \chi f^k, \quad \chi > 0 \tag{11.1}$$

where k is the trial index. Each measure of performance, y^k (such as peak velocity, accuracy, effort, or smoothness), is represented as a function of the system's current state z^k (i.e., a measure of a subject's current voluntary control) and of the amount of assistance f^k provided by the robot on the current trial modulated by χ. The motor recovery process, which represents the evolution of a subject's voluntary control, is described by

$$z^{k+1} = \vartheta z^k + \alpha e^k, \quad \vartheta, \alpha > 0 \tag{11.2}$$

where ϑ represents the retention rate, α the learning rate, and e^k the error term, which is an additional system input.

Casadio and Sanguineti (2012) used this model to describe the evolution of the average speed of movements performed during robot-assisted therapy with and without visual feedback. Despite its simplicity, the model was able to predict the experimentally observed trend in the average movement speed. In particular, the retention rate ϑ was able to accurately predict the score of the Fugl-Meyer test, a standard clinical assessment of performance, carried out by physiotherapists three months after the experiment with the robot! These promising results will likely motivate researchers to further develop subject-specific models of sensorimotor recovery, which may contribute to optimal recovery through the administration of appropriate clinical interventions.

11.8   Concurrent Force and Impedance Adaptation in Robots

New applications in service robotics, health care, and small-batch manufacturing require efficient interaction with unknown or changing environments, as well as with humans. These interactions may be unstable. For example, tools may slip if the direction of applied force is not properly aligned. Although humans can learn to perform such tasks with ease, robots generally lack the ability to adapt to instability because they have not been designed to deal with unstable interactions. However, recently the computational model of human motor adaptation presented in chapter 8 has been applied to robotic systems (Ganesh, Albu-Schaeffer, et al., 2010; Yang et al., 2011; Ganesh et al., 2012), and its capabilities have been illustrated in several applications.

If a task has predictable dynamics, the forces required to perform it can be learned. This observation is the essence of iterative control and adaptive control presented in chapter 6. However, learning an appropriate force is not relevant for dealing with unpredictable dynamics or instability. Success in unstable tasks requires a mechanism for learning suitable impedance. In chapter 8, we showed that humans are able to learn to perform tasks with appropriate forces and impedances in novel mechanical environments, using a strategy that appears to maintain a constant level of stability, and concurrently minimize error and effort (Franklin et al., 2008; Tee et al., 2010). Here we describe how a computational model that implements these features can be adapted as a biomimetic controller for robots and demonstrate its application in typical tasks performed with tools and in robot-human interaction.

Consider a robot mechanism with N actuated revolute joints, for which we want to specify the motor commands $\boldsymbol{\tau} \equiv (\tau_1,...,\tau_i,...,\tau_N)^T$. The control law from the computational model of equation (8.3) can then be adapted for the robot as

$$\boldsymbol{\tau} \equiv \mathbf{K}_0\boldsymbol{\varepsilon} + \boldsymbol{\tau}_{FF} + \mathbf{K}\boldsymbol{\varepsilon}, \quad \boldsymbol{\varepsilon} \equiv \mathbf{e} + \delta\dot{\mathbf{e}}, \quad \mathbf{e} = \mathbf{q}_u - \mathbf{q}, \quad \delta > 0 \qquad (11.3)$$

where $\boldsymbol{\varepsilon}$ is the *tracking error* relative to the planned trajectory $\mathbf{q}_u$ in joint space and $\mathbf{K}_0\boldsymbol{\varepsilon}$ is the stability margin used for unconstrained motion. When the robot interacts with the environment, the feedforward and feedback torque vectors, $\boldsymbol{\tau}_{FF}$ and $\mathbf{K}\boldsymbol{\varepsilon}$, are adapted to compensate for the interaction forces and impedance.

To provide a biomimetic controller for this learning, we first decompose the adaptation law of equation (8.5) as

$$\Delta \mathbf{u}^k \equiv \alpha[\boldsymbol{\varepsilon}^k]_+ + \alpha\chi[-\boldsymbol{\varepsilon}^k]_+ - \gamma\mathbf{1} = \frac{\alpha}{2}(1-\chi)\boldsymbol{\varepsilon}^k + \frac{\alpha}{2}(1+\chi)\mid\boldsymbol{\varepsilon}^k\mid-\gamma\mathbf{1} \tag{11.4}$$

where $\mid\boldsymbol{\varepsilon}\mid \equiv (\mid\varepsilon_1\mid,...,\mid\varepsilon_i\mid,...,\mid\varepsilon_M\mid)^T$ is defined component-wise. In this representation, the first term produces a force opposed to the error (i.e., compensates for systematic error), the second term in $\mid\boldsymbol{\varepsilon}\mid$ increases coactivation in response to deviation (i.e., increases stability), and the third term $-\gamma\mathbf{1}$ removes superfluous (co)activation, thus minimizing effort.

Examine the first term of equation (11.4), which produces a modification of reciprocal activation. Imagine that each joint is spanned by a pair of antagonistic muscles (neglecting the moment arms) then this term yields the adaptation of torque at this joint:

$$\Delta\boldsymbol{\tau}_{FF}^k \equiv \boldsymbol{\tau}_{FF}^{k+1} - \boldsymbol{\tau}_{FF}^k \equiv \alpha\boldsymbol{\varepsilon}^k \tag{11.5}$$

This equation is similar to the force regulation algorithms of nonlinear adaptive control, iterative learning control (described in chapter 6), and previous models of motor learning (Kawato, Furukawa & Suzuki, 1987; Sanner & Kosha, 1999). The other terms of equation (11.4) tune the coactivation in all antagonistic muscles groups; that is, our scheme extends these algorithms to simultaneous regulation of force and impedance. Assuming that stiffness is a linear function of tension, we have

$$\Delta K^k \equiv K^{k+1} - K^k \equiv \beta\mid\varepsilon^k\mid-\gamma, \; \beta,\gamma > 0 \tag{11.6}$$

In equations (11.5) and (11.6), k corresponds to either the time or the trial index. These equations can thus specify adaptation either at each time step or for the whole trajectory on a trial-by-trial basis. With these equations, the endpoint force and viscoelasticity are adapted to compensate for both internal and external forces and instability by limiting error and effort and ensuring a constant stability margin. The learning controller increases feedforward force and impedance as long as the error is large, that is, until the disturbing effect that results from the interaction with the environment is overcome. In the absence of a disturbance, the control reduces feedforward commands and impedance, retaining a small stability margin while maintaining a compliant interaction.

11.9   Robotic Implementation

The dynamic properties of this novel adaptive controller have been analyzed by Yang et al. (2011). Simulations exhibited an adaptive behavior similar to that of humans. Its ability to tune both impedance magnitude and geometry is illustrated in figure 11.6 on a 7DOF robot. The top left corner of figure 11.6 shows the robot in the configuration it attempts to maintain in the presence of disturbances generated by a human experimenter. The robot

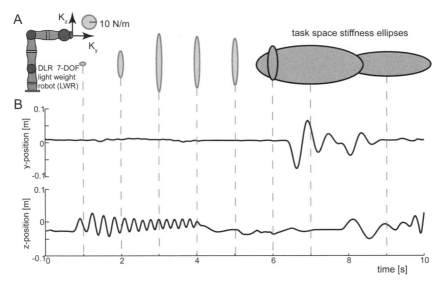

Figure 11.6
The endpoint stiffness of a multiple-degree-of-freedom robot (DLR 7DOF Light-Weight Robot) can be shaped through interaction. The biomimetic learning controller shapes the task stiffness in magnitude and direction while maintaining a posture (shown in the top-left corner) against disturbances applied at the end effector by a human. The projection of the translational task space stiffness matrix onto the yz plane is shown at different times.

reacts to a disturbance by increasing the task space stiffness along the direction of the disturbance, as humans do (Burdet et al., 2001). Stiffening occurs only in joints that are affected by the disturbance, so there is no stiffening, for example, of the proximal joints that have a large moment of inertia.

This biomimetic controller also adapts the force to the environment (figure 11.7). In this example, the feedforward command of a 1DOF robot is modified when a load is carried by the end effector. The desired position of the robot oscillates between 0 and 0.7 radians relative to the vertical. The experiment begins with an unknown load attached to the link, which causes a deviation from the reference trajectory that is reduced within five trials through adaptation of the feedforward torque. On the sixth trial, an extension spring is attached to the link that reduces the movement amplitude (gray region of figure 11.7A). However, the robot again learns the torque required to accomplish the task within nine movements. The spring is removed on the fifteenth trial (second white region in figure 11.7), causing an overshoot, which results in the robot readapting to torque levels similar to those before adding the spring. Figure 11.7C shows that a large error is accompanied by an increase in stiffness, which initially improves the control while the feedforward torque is being adapted. It then relaxes as the error decreases.

The versatility of the biomimetic controller in performing various interaction tasks has recently been illustrated by its application to typical tasks with tools such as drilling and

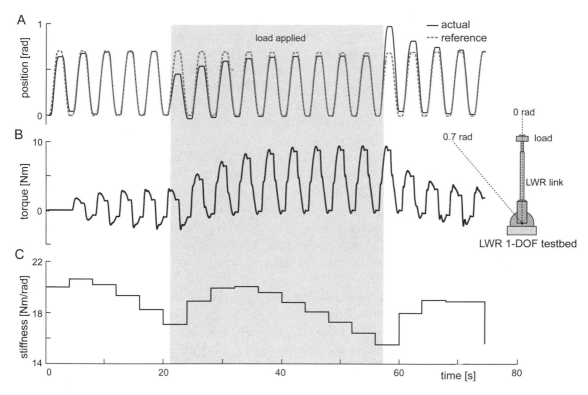

Figure 11.7
Adaptation of feedforward control using a single degree-of-freedom robot (DLR 1DOF Light-Weight Robot). Starting with an unknown load, the periodic robot motion (solid trace in [A]) is adapted to follow the reference (dashed trace in [A]). On addition (shaded region) and removal (subsequent white region) of a spring load, the error is compensated by the change of feedforward torque shown in (B). Impedance of the robot increases whenever there are novel (and thus unpredictable) dynamics but falls once the torque profile is appropriately adapted.

cutting, in which it exhibited superior performances to fixed-gain controllers during interaction with unknown materials (Ganesh et al., 2012). In summary, this controller can simultaneously adapt force and impedance in the presence of unknown dynamics, deal with unstable situations that are typical of tool use, and gradually acquire a desired stability margin. It does not require force sensing and generalizes to multiple movements (as described in section 8.4).

11.10   Humanlike Adaptation of Robotic Assistance for Active Learning

This controller can also be used to produce an intuitive adaptive human-robot interaction, such as that required for rehabilitation or teleoperation. Here we examine one example of using this controller for the adaptive control of rehabilitation robots. Although little is known about how robotics can best assist patients in recovering motor function after a stroke, clinical evidence has shown that one condition of success is the active participation of the patient (Hogan et al., 2006; Kahn et al., 2006), without which the patient fails to improve her or his motor function. Algorithms have therefore been proposed to adapt guidance stiffness (Fasoli, Krebs & Hogan, 2004) or the feedforward force provided by the robot (Wolbrecht et al., 2008) during training. The adaptive algorithm described in section 11.8 could be used to provide a gradual adaptation of trajectory, force, and impedance in a simple and automatic way.

To illustrate this, wrist flexion/extension training with the Bi-Manu-Track (Hesse et al., 2005) was simulated. With this robotic device, the nonaffected arm assists in oscillatory movements of the affected arm by providing force and guidance directly or mediated by a computer controlled torque actuator. In the subacute phase after a stroke, many patients cannot move the affected limb or at least lack control over the motion. Therefore, one strategy is to gradually adapt the trajectory amplitude, feedforward force, and guidance strength to the functional capacity of the affected limb using the adaptive learning algorithm.

This principle was incorporated into a simulation of impaired control of one limb. The torque applied by the robot to the affected limb assists the affected wrist to follow a trajectory provided by a healthy wrist, which is represented in this simulation as a minimal jerk trajectory (equation [5.38]). In the first iteration $t \in [0, T = 2s]$, the feedforward torque and stiffness are 0 (figure 11.8). The affected limb moves poorly on the first iteration, producing a large error. On the second iteration, the stiffness and feedforward torque provided by the robot increase to assist the wrist movement, which is assumed to improve gradually. We observe that as the position tracking error decreases, the robot feedforward torque and torque contributed by impedance decrease as well.

This simple simulation illustrates that the robot would provide only the minimal assistance required by the patient, who must therefore provide as much effort and perform to the best of his or her ability in order to succeed in the task. A similar scheme with adaptation of feedforward force only has been validated experimentally (Wolbrecht et al., 2008).

11.11   Summary and Conclusion

This chapter has presented several applications of the knowledge gained from studies of human neuromechanical control and motor learning to neurorehabilitation and to robotic control. We have seen that knowledge of neuromechanics can be used as a tool to improve neurorehabilitation. In particular, sensory feedback can be modulated to increase learning speed and retention. Although only proprioceptive and visual feedback has been discussed here, feedback from other sensory modalities such as auditory and haptic, EMG and EEG biofeedback, and electrical or magnetic stimulation of muscles, nerves, and neurons may be combined to promote adaptation. In general, ideas for new protocols can be tested in

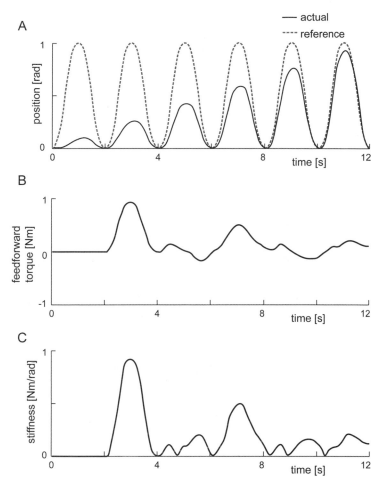

Figure 11.8
Adaptive assistance provided to a stroke-impaired wrist modified through humanlike learning by the robot controller. Assuming that the motor function is improving, the range of motion will increase. Assistance provided by the robot improves performance but decreases automatically when the impaired limb is able to produce the motion; that is, feedforward torque and stiffness provided by the robot decrease.

motor learning studies with healthy subjects and screened so that only promising approaches are then tested further with impaired subjects.

Obviously, the models presented in previous chapters consider only a few of the factors necessary to ensure effective neurorehabilitation. For instance, we have not considered spasticity, which is a major factor in neurological diseases such as cerebral palsy and stroke, or the typical attention deficits that accompany traumatic brain injury, or pathological tremor. Nevertheless, even though the application of computational principles of neuromechanics and motor learning to physical rehabilitation is in its infancy, it provides quantitative measures and modeling tools needed for a scientific approach to neurorehabilitation. Similar applications are starting to be developed in sport science, where computational models and robotic training can be used to optimize performance, and in prosthetics, where such knowledge can be used to design efficient neuroprostheses.

To control robots in physical interaction tasks, we have to solve the same problems that humans routinely solve efficiently without conscious effort. Therefore, it appears natural to examine "human solutions" to solve the problems we encounter in conceiving and programming robots. However, to our knowledge only a few biomimetic control models have so far been implemented on robotic platforms. One of these is impedance control, with promising implementations in prosthetics (Abul-Haj & Hogan, 1990), in the generation of "explosive" movements based on rapid release of elastic energy in tasks such as throwing (Braun, Howard & Vijayakumar, 2012) and in the control of mechanical interaction with the environment as was described previously. In this context, the computational model of chapter 8 gave rise to a novel adaptive algorithm for interactive control of robots with unknown dynamics and interaction with humans. We hope that this book will open new avenues for research on human motor control and lead to advances in related fields such as neurotechnology, physical rehabilitation and training, and robotics.

Appendix: Variable Definitions

Muscle

F_s: sarcomere force

x_s: sarcomere length

K_s: sarcomere stiffness

D_s: sarcomere coefficient of viscosity

F_t: tendon force

x_t: tendon length

K_t: tendon stiffness

D_t: tendon coefficient of viscosity

Muscle space

$\boldsymbol{\lambda} = (\lambda_1, \lambda_2, \ldots, \lambda_M)^T$

$\dot{\boldsymbol{\lambda}}$: muscle velocity

$\boldsymbol{\rho}$: moment arms

$\boldsymbol{\mu}$: muscle tensions

$\mathbf{J}_\mu(\boldsymbol{\rho})$: muscle Jacobian defined by $\dot{\boldsymbol{\lambda}} = \mathbf{J}_\mu(\mathbf{q})\,\dot{\mathbf{q}}$

$\mathbf{u}$: muscle activations

K_μ: muscle stiffness (single muscle)

D_μ: coefficient of muscle viscosity (single muscle)

$\mathbf{K}_\mu$: muscle stiffness matrix

$\mathbf{D}_\mu$: muscle viscosity matrix

Joint space

$\mathbf{q} = (q_1, q_1, \ldots, q_N)^T$: joint angles

$\dot{\mathbf{q}}$: angular velocity

$\ddot{\mathbf{q}}$: angular acceleration

$\boldsymbol{\tau}$: joint torques

$\mathbf{K} \equiv \left(-\dfrac{\partial \tau_i}{\partial q_j} - \sum_k \dfrac{\partial \tau_i}{\partial u_k} \dfrac{\partial u_k}{\partial q_j} \right)$: joint stiffness matrix

$\mathbf{D} \equiv \left(-\dfrac{\partial \tau_i}{\partial \dot{q}_j} - \sum_k \dfrac{\partial \tau_i}{\partial u_k} \dfrac{\partial u_k}{\partial \dot{q}_j} \right)$: joint viscosity matrix

$\mathbf{I} \equiv \left[-\dfrac{\partial \tau_i}{\partial \ddot{q}_j} \right]$: joint inertia matrix

Task space

$\mathbf{x} = (x_1, x_2, x_3)^T$: position

$\dot{\mathbf{x}}$: velocity

$\ddot{\mathbf{x}}$: acceleration

$\boldsymbol{\theta} = (\theta_1, \theta_2, \ldots, \theta_N)^T$: orientation

$\mathbf{J(q)}$: geometric Jacobian defined by $\dot{\mathbf{x}} = \mathbf{J(q)}\,\dot{\mathbf{q}}$

$\mathbf{J}^\dagger = \mathbf{J}^T (\mathbf{JJ}^T)^{-1}$: pseudo-inverse of the Jacobian

$\mathbf{F}$: Cartesian force

$\mathbf{K}_x \equiv \left(-\dfrac{\partial F_i}{\partial x_j} - \sum_k \dfrac{\partial F_i}{\partial u_k} \dfrac{\partial u_k}{\partial x_j} \right)$: endpoint stiffness matrix

$\mathbf{D}_x \equiv \left(-\dfrac{\partial F_i}{\partial \dot{x}_j} - \sum_k \dfrac{\partial F_i}{\partial u_k} \dfrac{\partial u_k}{\partial \dot{x}_j} \right)$: endpoint viscosity matrix

$\mathbf{I}_x \equiv \left(-\dfrac{\partial F_i}{\partial \ddot{x}_j} \right)$: endpoint inertia matrix

3-joint arm model

$(x, y, \theta)^T$: position and orientation in a plane

s: shoulder, e: elbow, w: wrist

$\mathbf{q} = (q_s, q_e, q_w)^T$: joint angles

$\mathbf{l} = (l_s, l_e, l_w)^T$: limb segment lengths

$\mathbf{m} = (m_s, m_e, m_w)^T$: masses of upper arm, forearm, hand

$\mathbf{lm} = (l_s m_s, l_e m_e, l_w m_w)^T$: distances to the respective center of mass

$\mathbf{I} = (I_s, I_e, I_w)^T$: moments of inertia

2-joint 6-muscle arm model

$\boldsymbol{\lambda} = (\lambda_{s+}, \lambda_{s-}, \lambda_{bs+}, \lambda_{bs-}, \lambda_{be+}, \lambda_{be-}, \lambda_{e+}, \lambda_{e-})^T$: muscles length

$\boldsymbol{\rho} = (\rho_{s+}, \rho_{s-}, \rho_{bs+}, \rho_{bs-}, \rho_{be+}, \rho_{be-}, \rho_{e+}, \rho_{e-})^T$: muscles moment arm

Rigid body model

$\boldsymbol{\tau} = \mathbf{H}(\mathbf{q})\ddot{\mathbf{q}} + \mathbf{C}(\mathbf{q}, \dot{\mathbf{q}})\dot{\mathbf{q}} + \mathbf{G}(\mathbf{q}) = \boldsymbol{\Psi}(\mathbf{q}, \dot{\mathbf{q}}, \ddot{\mathbf{q}})\,\mathbf{p}$: rigid body dynamics (in joint space)

$\boldsymbol{\tau}(\mathbf{p}) \equiv \boldsymbol{\Psi}\,\mathbf{p}$: linear form of dynamics (rigid body dynamics, synergies, force fields)

$\mathbf{p} = (p_1, \ldots, p_P)^T$: parameter vector

$\boldsymbol{\Psi}\mathbf{p} \equiv \mathbf{W}\boldsymbol{\Phi}$: linear neural network

$\boldsymbol{\Phi} = (\phi_1, \phi_2, \ldots, \phi_n)^T$: hidden layer of (artificial) neurons

Control

$\mathbf{q}_u(t), \; 0 \le t \le T$: undisturbed trajectory (human), desired trajectory (robot)

$\mathbf{F}_E$: force applied by the environment on the body

$\mathbf{e} = \mathbf{q}_u - \mathbf{q}$: position error

$\boldsymbol{\varepsilon} = \mathbf{e} + \kappa\dot{\mathbf{e}}$: tracking error

$\boldsymbol{\tau}_{FF}$: feedforward torque

$\boldsymbol{\tau}_{FB}$: feedback torque

ς: time delay

$\zeta = \dfrac{D}{2\sqrt{K\,I}}$: damping ratio

Motor adaptation

V: cost function to optimize

α: learning factor

β: coactivation learning factor

γ: forgetting factor

ϑ: retention factor

Stochastic optimal control

$\mathbf{z}(t)$: state space vector at time t

$\boldsymbol{\eta}$: motor noise

$E[\cdot]$: expected value (or mean value) of the stochastic function it is applied on

cost function to minimize:

$$V(\mathbf{u}) = E[h(\mathbf{z}(T_0))] + E\left[\int_0^T g(\mathbf{z}(t), \mathbf{u}(t), t)\, dt\right]$$

h: terminal or task cost

g: running cost

$g(\mathbf{u}) \equiv \mathbf{u}^T \mathbf{R}^T \mathbf{R} \mathbf{u}$: quadratic measure of effort

Linear discrete time model

$\mathbf{z}_{k+1} = \mathbf{A}_k \mathbf{z}_k + \mathbf{B}_k \mathbf{u}_k + \boldsymbol{\eta}_k$: system process

$\mathbf{y}_{k+1} = \mathbf{C}_k \mathbf{z}_k + \boldsymbol{\omega}_k$: observation process

$\boldsymbol{\omega}_k$: sensor noise

References

Abul-Haj, C. & Hogan, N. (1990). Functional assessment of control systems for cybernetic elbow prostheses: Application of the technique. *IEEE Transactions on Bio-Medical Engineering*, *37*(11), 1037–1047.

Adams, W. J., Graf, E. W. & Ernst, M. O. (2004). Experience can change the "light-from-above" prior. *Nature Neuroscience*, *7*(10), 1057–1058.

Akazawa, K., Milner, T. E. & Stein, R. B. (1983). Modulation of reflex EMG and stiffness in response to stretch of human finger muscle. *Journal of Neurophysiology*, *49*(1), 16–27.

Albus, J. (1971). A theory of cerebellar function. *Mathematical Biosciences*, *10*, 25–61.

Andersen, R. A. & Cui, H. (2009). Intention, action planning, and decision making in parietal-frontal circuits. *Neuron*, *63*(5), 568–583.

Ariff, G., Donchin, O., Nanayakkara, T. & Shadmehr, R. (2002). A real-time state predictor in motor control: Study of saccadic eye movements during unseen reaching movements. *Journal of Neuroscience*, *22*(17), 7721–7729.

Atkeson, C. G. & Hollerbach, J. M. (1985). Kinematic features of unrestrained vertical arm movements. *Journal of Neuroscience*, *5*(9), 2318–2330.

Balasubramanian, S., Colombo, R., Sterpi, I., Sanguineti, V. & Burdet, E. (2012). Robotic assessment of upper-limb motor function after stroke: A review. *American Journal of Physical Medicine & Rehabilitation*, *91*(11), 255–269.

Balasubramanian, S., Klein, J. & Burdet, E. (2010). Robot-assisted rehabilitation of hand function. *Current Opinion in Neurology*, *23*(6), 661–670.

Bear, M. F., Connors, B. W. & Paradiso, M. A. (2007). *Neuroscience: Exploring the brain*. Baltimore: Lippincott Williams and Wilkins.

Bennett, D. J., Gorassini, M. & Prochazka, A. (1994). Catching a ball: Contributions of intrinsic muscle stiffness, reflexes, and higher order responses. *Canadian Journal of Physiology and Pharmacology*, *72*(5), 525–534.

Bernier, P.-M., Chua, R., Bard, C. & Franks, I. M. (2006). Updating of an internal model without proprioception: A deafferentation study. *Neuroreport*, *17*(13), 1421–1425.

Beurze, S. M., Toni, I., Pisella, L. & Medendorp, W. P. (2010). Reference frames for reach planning in human frontoparietal cortex. *Journal of Neurophysiology*, *104*(3), 1736–1745.

Bien, Z. & Xu, J. X. (1998). *Iterative learning control: Analysis, design, integration and applications*. Boston: Kluwer Academic.

Biess, A., Flash, T. & Liebermann, D. G. (2011). Riemannian geometric approach to human arm dynamics, movement optimization and invariance. *Physical Review E: Statistical, Nonlinear, and Soft Matter Physics*, *83*, 031927.

Biess, A., Liebermann, D. G. & Flash, T. (2007). A computational model for redundant human three dimensional pointing movements: Integration of independent spatial and temporal motor plans simplifies movement dynamics. *Journal of Neuroscience*, *27*(48), 13045–13064.

Bizzi, E., Cheung, V. C. K., d'Avella, A., Saltiel, P. & Tresch, M. C. (2008). Combining modules for movement. *Brain Research Reviews*, *57*(1), 125–133.

Blakemore, S. J., Frith, C. D. & Wolpert, D. M. (1999). Spatio-temporal prediction modulates the perception of self-produced stimuli. *Journal of Cognitive Neuroscience, 11*(5), 551–559.

Bosnell, R. A., Kincses, T., Stagg, C. J., Tomassini, V., Kischka, U., Jbabdi, S., et al. (2011). Motor practice promotes increased activity in brain regions structurally disconnected after subcortical stroke. *Neurorehabilitation and Neural Repair, 25*(7), 607–616.

Bouisset, F. & Lestienne, F. (1974). The organisation of simple voluntary movement as analysed from its kinematic properties. *Brain Research, 71*(2–3), 451–457.

Brashers-Krug, T., Shadmehr, R. & Bizzi, E. (1996). Consolidation in human motor memory. *Nature, 382*(6588), 252–255.

Braun, D., Howard, M. & Vijayakumar, S. (2012). Optimal variable stiffness control: Formulation and application to explosive movement tasks. *Autonomous Robots, 33*(3), 237–253.

Brooks, S. V. & Faulkner, J. A. (1994). Skeletal muscle weakness in old age: Underlying mechanisms. *Medicine and Science in Sports and Exercise, 26*(4), 432–439.

Brown, S. H. & Cooke, J. D. (1990). Movement-related phasic muscle activation I. Relations with temporal profile of movement. *Journal of Neurophysiology, 63*(3), 455–464.

Buneo, C. A., Jarvis, M. R., Batista, A. P. & Andersen, R. A. (2002). Direct visuomotor transformations for reaching. *Nature, 416*(6881), 632–636.

Burdet, E. & Codourey, A. (1998). Evaluation of parametric and nonparametric nonlinear adaptive controllers. *Robotica, 16*(1), 59–73.

Burdet, E., Codourey, A. & Rey, L. (1998). Experimental evaluation of nonlinear adaptive controllers. *IEEE Control Systems Magazine, 18*(2), 39–47.

Burdet, E., Ganesh, G., Yang, C. & Albu-Schaeffer, A. (2013). Learning interaction force, impedance and trajectory: By humans, for robots. O. Khatib, V. Kumar, & G. Sukhatme (Eds.), *The 12th International Symposium on Experimental Robotics. Springer Tracts in Advanced Robotics, 79*.

Burdet, E., Honegger, M. & Codourey, A. (2000). Controllers with desired dynamics and their implementation on a 6 DOF parallel manipulator. *IROS 2000 Proceedings IEEE/RSJ International Conference on Robotics and Intelligent Systems (IROS), 1*, 39–45.

Burdet, E. & Milner, T. E. (1998). Quantization of human motions and learning of accurate movements. *Biological Cybernetics, 78*(4), 307–318.

Burdet, E., Osu, R., Franklin, D., Yoshioka, T., Milner, T. & Kawato, M. (2000). A method for measuring endpoint stiffness during multi-joint arm movements. *Journal of Biomechanics, 33*(12), 1705–1709.

Burdet, E., Osu, R., Franklin, D., Milner, T. & Kawato, M. (2001). The central nervous system stabilizes unstable dynamics by learning optimal impedance. *Nature, 414*(6862), 446–449.

Burdet, E., Tee, K. P., Mareels, I., Milner, T. E., Chew, C. M., Franklin, D. W., et al. (2006). Stability and motor adaptation in human arm movements. *Biological Cybernetics, 94*(1), 20–32.

Burgar, C. B., Lum, P. S., Shor, P. C. & Van der Loos, M. H. F. (2000). Development of robots for rehabilitation therapy: The Palo Alto VA/Stanford experience. *Journal of Rehabilitation Research and Development, 37*(6), 663–673.

Caithness, G., Osu, R., Bays, P., Chase, H., Klassen, J., Kawato, M., et al. (2004). Failure to consolidate the consolidation theory of learning for sensorimotor adaptation tasks. *Journal of Neuroscience, 24*(40), 8662–8671.

Campbell, K. S. (2010). Short-range properties of skeletal and cardiac muscles. *Advances in Experimental Medicine and Biology, 682*, 223–246.

Capaday, C., Forget, R. & Milner, T. E. (1994). A reexamination of the effects of instruction on the long latency stretch reflex response of the flexor pollicis longus muscle. *Experimental Brain Research, 100*(3), 515–521.

Casadio, M. & Sanguineti, V. (2012). Learning, retention, and slacking: A model of the dynamics of recovery in robot therapy. *IEEE Transactions on Neural Systems and Rehabilitation Engineering, 20*(3), 286–296.

Churchland, M. M., Afshar, A. & Shenoy, K. V. (2006). A central source of movement variability. *Neuron, 52*(6), 1085–1096.

Codourey, A. & Burdet, E. (1997). A body-oriented method for finding a linear form of the dynamic equation of fully parallel robots. *ICRA 1997 Proceedings IEEE International Conference on Robotics and Automation (ICRA)*, 2, 1612–1618.

Colgate, J. E. & Hogan, N. (1988). Robust control of dynamically interacting systems. *International Journal of Control*, 48(1), 65–88.

Conditt, M., Gandolfo, F. & Mussa-Ivaldi, F. (1997). The motor system does not learn the dynamics of the arm by rote memorization of past experience. *Journal of Neurophysiology*, 78(1), 554–560.

Conditt, M. A. & Mussa-Ivaldi, F. A. (1999). Central representation of time during motor learning. *Proceedings of the National Academy of Sciences of the United States of America*, 96(20), 11625–11630.

Craig, J. J. (2005). *Introduction to robotics: Mechanics and control*. Upper Saddle River, NJ: Prentice Hall.

Crevecoeur, F., Kurtzer, I. & Scott, S. H. (2012). Fast corrective responses are evoked by perturbations approaching the natural variability of posture and movement tasks. *Journal of Neurophysiology*, 107(10), 2821–2832.

Criscimagna-Hemminger, S. E., Bastian, A. J. & Shadmehr, R. (2010). Size of error affects cerebellar contributions to motor learning. *Journal of Neurophysiology*, 103(4), 2275–2284.

Cruse, H., Brüwer, M. & Dean, J. (1993). Control of three- and four-joint arm movement: Strategies for a manipulator with redundant degrees of freedom. *Journal of Motor Behavior*, 25(3), 131–139.

Cruse, H., Wischmeyer, E., Brüwer, M., Brockfeld, P. & Dress, A. (1990). On the cost functions for the control of the human arm movement. *Biological Cybernetics*, 62(6), 519–528.

D'Avella, A., Fernandez, L., Portone, A. & Lacquaniti, F. (2008). Modulation of phasic and tonic muscle synergies with reaching direction and speed. *Journal of Neurophysiology*, 100(3), 1433–1454.

D'Avella, A., Portone, A., Fernandez, L. & Lacquaniti, F. (2006). Control of fast-reaching movements by muscle synergy combinations. *Journal of Neuroscience*, 26(30), 7791–7810.

Dapena, J. (1980). Mechanics of rotation in the Fosbury-flop. *Medicine and Science in Sports and Exercise*, 12(1), 45–53.

Darainy, M., Malfait, N., Gribble, P. L., Towhidkhah, F. & Ostry, D. J. (2004). Learning to control arm stiffness under static conditions. *Journal of Neurophysiology*, 92(6), 3344–3350.

Day, B. & Lyon, I. (2000). Voluntary modification of automatic arm movements evoked by motion of a visual target. *Experimental Brain Research*, 130(2), 159–168.

Day, B. L., Riescher, H., Struppler, A., Rothwell, J. C. & Marsden, C. D. (1991). Changes in the response to magnetic and electrical stimulation of the motor cortex following muscle stretch in man. *Journal of Physiology*, 433, 41–57.

Day, B. L., Thompson, P. D., Harding, A. D. & Marsden, C. D. (1998). Influence of vision on upper limb reaching movements in patients with cerebellar ataxia. *Brain*, 121(Pt 2), 357–372.

De Serres, S. J. & Milner, T. E. (1991). Wrist muscle activation patterns and stiffness associated with stable and unstable mechanical loads. *Experimental Brain Research*, 86(2), 451–458.

Dean, P. & Porrill, J. (2011). Evaluating the adaptive filter model of the cerebellum. *Journal of Physiology*, 589(Pt 14), 3459–3470.

DeWit, C. C., Bastin, G. & Siciliano, B. (1996). *Theory of robot control*. Berlin: Springer.

Diedrichsen, J. (2007). Optimal task-dependent changes of bimanual feedback control and adaptation. *Current Biology*, 17(19), 1675–1679.

Dimitriou, M. & Edin, B. B. (2010). Human muscle spindles act as forward sensory models. *Current Biology*, 20(19), 1763–1767.

Dimitriou, M., Franklin, D. W. & Wolpert, D. M. (2012). Task-dependent coordination of rapid bimanual motor responses. *Journal of Neurophysiology*, 107(3), 890–901.

DiZio, P. & Lackner, J. R. (2000). Congenitally blind individuals rapidly adapt to coriolis force perturbations of their reaching movements. *Journal of Neurophysiology*, 84(4), 2175–2180.

Doemges, F. & Rack, P. M. (1992). Changes in the stretch reflex of the human first dorsal interosseous muscle during different tasks. *Journal of Physiology*, 447, 563–573.

Donchin, O., Francis, J. T. & Shadmehr, R. (2003). Quantifying generalization from trial-by-trial behavior of adaptive systems that learn with basis functions: Theory and experiments in human motor control. *Journal of Neuroscience*, *23*(27), 9032–9045.

Doya, K. (2000). Reinforcement learning in continuous time and space. *Neural Computation*, *12*(1), 219–245.

Dulhunty, A. F. (2006). Excitation-contraction coupling from the 1950s into the new millennium. *Clinical and Experimental Pharmacology & Physiology*, *33*(9), 763–772.

Edman, A. K. (2010). Contractile performance of striated muscle. *Advances in Experimental Medicine and Biology*, *682*, 7–40.

Ellis, M. D., Sukal-Moulton, T., & Dewald, J. P. A. (2009). Progressive shoulder abduction loading is a crucial element of arm rehabilitation in chronic stroke. *Neurorehabilitation and Neural Repair*, *23*(8), 862–869.

Emken, J. L. & Reinkensmeyer, D. J. (2005). Robot- enhanced motor learning: Accelerating internal model formation during locomotion by transient dynamic amplification. *IEEE Transactions on Neural Systems and Rehabilitation Engineering*, *13*(1), 33–39.

Enoka, R. M. & Duchateau, J. (2008). Muscle fatigue: What, why and how it influences muscle function. *Journal of Physiology*, *586*(1), 11–23.

Enoka, R. M. & Stuart, D. G. (1992). Neurobiology of muscle fatigue. *Journal of Applied Physiology*, *72*(5), 1631–1648.

Ernst, M. O. & Banks, M. S. (2002). Humans integrate visual and haptic information in a statistically optimal fashion. *Nature*, *415*(6870), 429–433.

Ernst, M. O. & Bülthoff, H. H. (2004). Merging the senses into a robust percept. *Trends in Cognitive Sciences*, *8*(4), 162–169.

Eyre, J. A., Miller, S. & Ramesh, V. (1991). Constancy of central conduction delays during development in man: Investigation of motor and somatosensory pathways. *Journal of Physiology*, *434*, 441–452.

Faisal, A. A., Selen, L. P. J. & Wolpert, D. M. (2008). Noise in the nervous system. *Nature Reviews Neuroscience*, *9*(4), 292–303.

Fasoli, S., Krebs, H. & Hogan, N. (2004). Robotic technology and stroke rehabilitation: Translating research into practice. *Topics in Stroke Rehabilitation*, *11*(4), 11–19.

Featherstone, R. (2008). *Rigid body dynamics algorithms*. New York: Springer. (See also code at http://royfeatherstone.org/spatial/v2/index.html.)

Firouzimehr, Z. (2011). The role of muscle cocontraction in motor learning. M.Sc. Thesis, McGill University.

Flanagan, J. R., Nakano, E., Imamizu, H., Osu, R., Yoshioka, T. & Kawato, M. (1999). Composition and decomposition of internal models in motor learning under altered kinematic and dynamic environments. *Journal of Neuroscience*, *19*(20), RC34.

Flanders, M., Tillery, S. I. & Soechting, J. F. (1992). Early stages in a sensorimotor transformation. *Behavioral and Brain Sciences*, *15*, 309–362.

Flash, T. & Henis, E. (1995). Arm trajectory modifications during reaching towards visual targets. *Journal of Cognitive Neuroscience*, *3*(3), 220–230.

Flash, T. & Hogan, N. (1985). The coordination of arm movements: An experimentally confirmed mathematical model. *Journal of Neuroscience*, *5*(7), 1688–1703.

Franklin, D. W., Burdet, E., Osu, R., Kawato, M. & Milner, T. E. (2003). Functional significance of stiffness in adaptation of multijoint arm movements to stable and unstable dynamics. *Experimental Brain Research*, *151*(2), 145–157.

Franklin, D. W., Burdet, E., Tee, K. P., Osu, R., Chew, C.-M., Milner, T. E., et al. (2008). CNS learns stable, accurate, and efficient movements using a simple algorithm. *Journal of Neuroscience*, *28*(44), 11165–11173.

Franklin, D. W., Liaw, G., Milner, T. E., Osu, R., Burdet, E. & Kawato, M. (2007). Endpoint stiffness of the arm is directionally tuned to instability in the environment. *Journal of Neuroscience*, *27*(29), 7705–7716.

Franklin, D. W., Osu, R., Burdet, E., Kawato, M. & Milner, T. E. (2003). Adaptation to stable and unstable dynamics achieved by combined impedance control and inverse dynamics model. *Journal of Neurophysiology*, *90*(5), 3270–3282.

Franklin, D. W., So, U., Burdet, E. & Kawato, M. (2007). Visual feedback is not necessary for the learning of novel dynamics. *PLoS ONE, 2*(12), e1336. doi:10.1371/journal.pone.0001336.

Franklin, D. W., So, U., Kawato, M. & Milner, T. E. (2004). Impedance control balances stability with metabolically costly muscle activation. *Journal of Neurophysiology, 92*(5), 3097–3105.

Franklin, D. W. & Wolpert, D. M. (2008). Specificity of reflex adaptation for task-relevant variability. *Journal of Neuroscience, 28*(52), 14165–14175.

Franklin, D. W. & Wolpert, D. M. (2011). Computational mechanisms of sensorimotor control. *Neuron, 72*(3), 425–442.

Franklin, S., Wolpert, D. M. & Franklin, D. W. (2012). Visuomotor feedback gains upregulate during the learning of novel dynamics. *Journal of Neurophysiology, 108*(2), 467–478.

Freund, H.-J. & Büdingen, H. J. (1978). The relationship between speed and amplitude of the fastest voluntary contractions of human arm muscles. *Experimental Brain Research, 31*(1), 1–12.

Galea, J. M., Vazquez, A., Pasricha, N., Orban de Xivry, J. J. & Celnik, P. (2011). Dissociating the roles of the cerebellum and motor cortex during adaptive learning: The motor cortex retains what the cerebellum learns. *Cerebral Cortex, 21*(8): 1761–1770.

Gandevia, S. C. (2011). Anesthesia: Roles for afferent signals and motor commands. *Comprehensive Physiology*, 128–172. doi:10.1002/cphy.cp120104.

Gandolfo, F., Mussa-Ivaldi, F. A. & Bizzi, E. (1996). Motor learning by field approximation. *Proceedings of the National Academy of Sciences of the United States of America, 93*(9), 3843–3846.

Ganesh, G., Albu-Schaeffer, A., Haruno, M., Kawato, M. & Burdet, E. (2010). Biomimetic motor behavior for simultaneous adaptation of force, impedance and trajectory in interaction tasks. *Proceedings ICRA 2010 IEEE International Conference on Robotics and Automation*, 2705–2711.

Ganesh, G., Haruno, H., Kawato, M. & Burdet, E. (2010). Motor memory and local minimization of error and effort, not global optimization, determine motor behavior. *Journal of Neurophysiology, 104*(1), 382–390.

Ganesh, G., Jarrasse, N., Haddadin, S., Albu-Schaeffer, A. & Burdet, E. (2012). A versatile biomimetic controller for contact tooling and tactile exploration. *Proceedings ICRA 2012 IEEE International Conference on Robotics and Automation*, 3329–3334.

Gans, C. & Gaunt, A. S. (1991). Muscle architecture in relation to function. *Journal of Biomechanics, 24*(Suppl. 1), 53–65.

Gao, Z., van Beugen, B. J. & De Zeeuw, C. I. (2012). Distributed synergistic plasticity and cerebellar learning. *Nature Reviews Neuroscience, 13*(9), 619–635.

Geeves, M. A. & Holmes, K. C. (1999). Structural mechanism of muscle contraction. *Annual Review of Biochemistry, 68*, 687–728.

Georgopoulos, A. P., Kalaska, J. F., Caminiti, R. & Massey, J. T. (1982). On the relations between the direction of two-dimensional arm movements and cell discharge in primate motor cortex. *Journal of Neuroscience, 2*(11), 1527–1537.

Georgopoulos, A. P., Kalaska, J. F. & Massey, J. T. (1981). Spatial trajectories and reaction times of aimed movements: Effects of practice, uncertainty, and change in target location. *Journal of Neurophysiology, 46*(4), 725–743.

Ghez, C. & Gordon, J. (1987). Trajectory control in targeted force impulses. I. Role of opposing muscles. *Experimental Brain Research, 67*(2), 225–240.

Ghez, C., Gordon, J., & Ghilardi, M. F. (1995). Impairments of reaching movements in patients without proprioception. II. Effects of visual information on accuracy. *Journal of Neurophysiology, 73*(1), 361–372.

Gomi, H., Abekawa, N. & Nishida, S. (2006). Spatiotemporal tuning of rapid interactions between visual-motion analysis and reaching movement. *Journal of Neuroscience, 26*(20), 5301–5308.

Gomi, H. & Kawato, M. (1997). Human arm stiffness and equilibrium-point trajectory during multi-joint movement. *Biological Cybernetics, 76*(3), 163–171.

Gomi, H. & Osu, R. (1998). Task-dependent viscoelasticity of human multijoint arm and its spatial characteristics for interaction with environments. *Journal of Neuroscience, 18*(21), 8965–8978.

Gonzalez Castro, L. N., Monsen, C. B. & Smith, M. A. (2011). The binding of learning to action in motor adaptation. *PLoS Computational Biology*, *7*(6), e1002052.

Goodbody, S. & Wolpert, D. (1998). Temporal and amplitude generalization in motor learning. *Journal of Neurophysiology*, *79*(4), 1825–1838.

Gordon, J., Ghilardi, M. F., Cooper, S. E. & Ghez, C. (1994). Accuracy of planar reaching movements. II. Systematic extent errors resulting from inertial anisotropy. *Experimental Brain Research*, *99*(1), 112–130.

Gordon, J., Ghilardi, M. F. & Ghez, C. (1995). Impairments of reaching movements in patients without proprioception. I. Spatial errors. *Journal of Neurophysiology*, *73*(1), 347–360.

Gordon, A. M., Huxley, A. F. & Julian, F. J. (1966). The variation in isometric tension with sarcomere length in vertebrate muscle fibres. *Journal of Physiology*, *184*(1), 170–192.

Gottlieb, G. L., Song, Q., Hong, D. & Corcos, D. M. (1996). Coordinating two degrees of freedom during human arm movement: Load and speed invariance of relative joint torques. *Journal of Neurophysiology*, *76*(5), 3196–3206.

Grey, M. J., Nielsen, J. B., Mazzaro, N. & Sinkjaer, T. (2007). Positive force feedback in human walking. *Journal of Physiology*, *581*(Pt 1), 99–105.

Gribble, P. L., Mullin, L. I., Cothros, N. & Mattar, A. (2003). Role of cocontraction in arm movement accuracy. *Journal of Neurophysiology*, *89*(5), 2396–2405.

Halsband, U. & Passingham, R. E. (1985). Premotor cortex and the conditions for movement in monkeys (Macaca fascicularis). *Behavioural Brain Research*, *18*(3), 269–277.

Hamilton, A. F., Jones, K. E. & Wolpert, D. M. (2004). The scaling of motor noise with muscle strength and motor unit number in humans. *Experimental Brain Research*, *157*(4), 417–430.

Harris, C. M. & Wolpert, D. M. (1998). Signal-dependent noise determines motor planning. *Nature*, *394*(6695), 780–784.

Haruno, M., Ganesh, G., Burdet, E. & Kawato, M. (2012). Distinct neural correlates of reciprocal- and co-activation of muscles in dorsal and ventral premotor cortices. *Journal of Neurophysiology*, *107*(1), 126–133.

Heckman, C. J. & Enoka, R. M. (2012). Motor unit. *Comprehensive Physiology*, *2*, 2629–2682.

Heller, A., Wade, D. T., Wood, V. A., Sunderland, A., Hewer, R. L. & Ward, E. (1987). Arm function after stroke: Measurement and recovery over the first three months. *Journal of Neurology, Neurosurgery, and Psychiatry*, *50*(6), 714–719.

Herbert, R. D. & Gandevia, S. C. (1995). Changes in pennation with joint angle and muscle torque: In vivo measurements in human brachialis muscle. *Journal of Physiology*, *484*(Pt 2), 523–532.

Herter, T. M., Kurtzer, I., Cabel, D. W., Haunts, K. A. & Scott, S. H. (2007). Characterization of torque-related activity in primary motor cortex during a multijoint postural task. *Journal of Neurophysiology*, *97*(4), 2887–2899.

Herzog, W., Joumaa, V. & Leonard, T. R. (2010). The force-length relationship of mechanically isolated sarcomeres. *Advances in Experimental Medicine and Biology*, *682*, 141–161.

Hesse, S., Werner, C., Pohl, M., Rueckriem, S., Mehrholz, J. & Lingnau, M. L. (2005). Computerized arm training improves the motor control of the severely affected arm after stroke a single-blinded randomized trial in two centers. *Stroke*, *36*(9), 1960–1966.

Hoffman, D. S. & Strick, P. L. (1990). Step-tracking movements of the wrist in humans. II. EMG analysis. *Journal of Neuroscience*, *10*(1), 142–152.

Hogan, N. (1984). Adaptive control of mechanical impedance by coactivation of antagonist muscles. *IEEE Transactions on Automatic Control*, *29*(8), 681–690.

Hogan, N. (1985). The mechanics of multi-joint posture and movement control. *Biological Cybernetics*, *52*(5), 315–331.

Hogan, N. (1990). Mechanical impedance of single-and multi-articular systems. Multiple muscle systems biomechanics and movement. In J. M. Winters & S. Woo (Eds.), *L-Y. Multiple muscle systems: Biomechanics and movement organization* (pp. 149–164). New York: Springer.

Hogan, N., Krebs, H. I., Rohrer, B., Palazzolo, J., Dipietro, L., Stein, J., et al. (2006). Motions or muscles? Some behavioral factors underlying robotic assistance of motor recovery. *Journal of Rehabilitation Research and Development*, *43*(5), 605–618.

Hollander, M., Koudstaal, P. J., Bots, M. L., Grobbee, D. E., Hofman, A. & Breteler, M. M. B. (2003). Incidence, risk, and case fatality of first ever stroke in the elderly population. The Rotterdam Study. *Journal of Neurology, Neurosurgery, and Psychiatry*, *74*(3), 317–321.

Hollerbach, M. J. & Flash, T. (1982). Dynamic interactions between limb segments during planar arm movement. *Biological Cybernetics*, *44*(1), 67–77.

Horak, F. B. (2009). Postural control. In M. D. Binder, N. Hirokawa, & U. Windhorst (Eds.), *Encyclopedia of neuroscience* (pp. 3212–3219). Berlin: Springer.

Howard, I. S., Ingram, J. N., Körding, K. P. & Wolpert, D. M. (2009). Statistics of natural movements are reflected in motor errors. *Journal of Neurophysiology*, *102*(3), 1902–1910.

Howard, I. S., Ingram, J. N. & Wolpert, D. M. (2008). Composition and decomposition in bimanual dynamic learning. *Journal of Neuroscience*, *28*(42), 10531–10540.

Howard, I. S., Ingram, J. N. & Wolpert, D. M. (2010). Context-dependent partitioning of motor learning in bimanual movements. *Journal of Neurophysiology*, *104*(4), 2082–2091.

Howard, I. S., Ingram, J. N. & Wolpert, D. M. (2011). Separate representations of dynamics in rhythmic and discrete movements: Evidence from motor learning. *Journal of Neurophysiology*, *105*(4), 1722–1731.

Hu, X., Murray, W. M., & Perreault, E. J. (2012). Biomechanical constraints on the feedforward regulation of endpoint stiffness. *Journal of Neurophysiology*, *108*(8), 2083–2091.

Huang, F. C. & Patton, J. L. (2012). Augmented dynamics and motor exploration as training for stroke. *IEEE Transactions on Bio-Medical Engineering*. doi:10.1109/TBME.2012.2192116.

Huang, V. S. & Shadmehr, R. (2009). Persistence of motor memories reflects statistics of the learning event. *Journal of Neurophysiology*, *102*(2), 931–940.

Hunter, I. W. & Kearney, R. E. (1982). Dynamics of human ankle stiffness: Variation with mean ankle torque. *Journal of Biomechanics*, *15*(10), 747–752.

Huxley, A. F. (1957). Muscle structure and theories of contraction. *Progress in Biophysics and Biophysical Chemistry*, *7*, 255–318.

Huxley, H. E. (1957). The double array of filaments in cross-striated muscle. *Journal of Biophysical and Biochemical Cytology*, *3*(5), 631–648.

Ichbiah, D. (2005). *Robots: From science fiction to technological revolution*. New York: Harry N. Abrams.

Imamizu, H., Kuroda, T., Miyauchi, S., Yoshioka, T. & Kawato, M. (2003). Modular organization of internal models of tools in the human cerebellum. *Proceedings of the National Academy of Sciences of the United States of America*, *100*(9), 5461–5466.

Imamizu, H., Miyauchi, S., Tamada, T., Sasaki, Y. Takino, R., Pütz, B., et al. (2000). Human cerebellar activity reflecting an acquired internal model of a new tool. *Nature*, *403*(6766), 192–195.

Ito, M. (1984). *The cerebellum and neural motor control*. New York: Raven Press.

Ito, M., Kawakami, Y., Ichinose, Y., Fukashiro, S., & Fukunaga, T. (1998). Nonisometric behavior of fascicles during isometric contractions of a human muscle. *Journal of Applied Physiology*, *85*(4), 1230–1235.

Ito, T., Murano, E. Z. & Gomi, H. (2004). Fast force-generation dynamics of human articulatory muscles. *Journal of Applied Physiology*, *96*(6), 2318–2324.

Iwamura, Y. (2009). Tactile senses—touch. In M. D. Binder, N. Hirokawa, & U. Windhorst (Eds.), *Encyclopedia of neuroscience* (pp. 4005–4009). Berlin: Springer.

Izawa, J., Rane, T., Donchin, O. & Shadmehr, R. (2008). Motor adaptation as a process of reoptimization. *Journal of Neuroscience*, *28*(11), 2883–2891.

Jacks, A., Prochazka, A. & Trend, P. S. J. (1988). Instability in human forearm movements studied with feedback-controlled electrical stimulation of muscles. *Journal of Physiology*, *402*, 443–461.

Jackson, S. R. & Husain, M. (1996). Visuomotor functions of the lateral pre-motor cortex. *Current Opinion in Neurobiology*, *6*(6), 788–795.

Jacobs, R. A. R. (1999). Optimal integration of texture and motion cues to depth. *Vision Research*, *39*(21), 3621–3629.

Jaeger, R. J., Gottlieb, G. L. & Agarwal, G. C. (1982). Myoelectric responses at flexors and extensors of human wrist to step torque perturbations. *Journal of Neurophysiology*, *48*(2), 388–402.

Jami, L. (1992). Golgi tendon organs in mammalian skeletal muscle: Functional properties and central actions. *Physiological Reviews*, *72*(3), 623–666.

Johansson, R. S. & Flanagan, J. R. (2009). Coding and use of tactile signals from the fingertips in object manipulation tasks. *Nature Reviews Neuroscience*, *10*(5), 345–359.

Johansson, R. S. & Westling, G. (1988). Programmed and triggered actions to rapid load changes during precision grip. *Experimental Brain Research*, *71*(1), 72–86.

Johansson, R. S., Westling, G., Bäckström, A. & Flanagan, J. R. (2001). Eye-hand coordination in object manipulation. *Journal of Neuroscience*, *21*(17), 6917–6932.

Jones, K. E., Hamilton, A. F. & Wolpert, D. M. (2002). Sources of signal-dependent noise during isometric force production. *Journal of Neurophysiology*, *88*(3), 1533–1544.

Jubrias, S. A., Odderson, I. R., Esselman, P. C. & Conley, K. E. (1997). Decline in isokinetic force with age: Muscle cross-sectional area and specific force. *Pflügers Archiv: European Journal of Physiology*, *434*(3), 246–253.

Kadiallah, A., Franklin, D. W. & Burdet, E. (2012). Generalization in adaptation to stable and unstable dynamics. *PLoS ONE*, *7*(10), e45075. doi:10.1371/journal.pone.0045075.

Kadiallah, A., Liaw, G., Kawato, M., Franklin, D. W. & Burdet, E. (2011). Impedance control is selectively tuned to multiple directions of movement. *Journal of Neurophysiology*, *106*(5), 2737–2748.

Kahn, L., Lum, P. S., Rymer, W. Z. & Reinkensmeyer, D. J. (2006). Robot-assisted movement training for the stroke-impaired arm: Does it matter what the robot does? *Journal of Rehabilitation Research and Development*, *43*(5), 619–630.

Kamper, D. G. & Rymer, W. Z. (2000). Quantitative features of the stretch response of extrinsic finger muscles in hemiparetic stroke. *Muscle & Nerve*, *23*(6), 954–961.

Kandel, E. R., Schwarz, J. H.,& Jessell, T. M. (2000). *Principles of neural science*. New York: McGraw-Hill.

Kargo, W. J. & Giszter, S. F. (2008). Individual premotor drive pulses, not time-varying synergies, are the units of adjustment for limb trajectories constructed in spinal cord. *Journal of Neuroscience*, *28*(10), 2409–2425.

Karniel, A. & Mussa-Ivaldi, F. A. (2003). Sequence, time, or state representation: How does the motor control system adapt to variable environments? *Biological Cybernetics*, *89*(1), 10–21.

Kawato, M. (1999). Internal models for motor control and trajectory planning. *Current Opinion in Neurobiology*, *9*(6), 718–727.

Kawato, M., Furukawa, K., & Suzuki, R. (1987). A hierarchical neural-network model for control and learning of voluntary movement. *Biological Cybernetics*, *57*(3), 169–185.

Kersten, D. D. & Yuille, A. A. (2003). Bayesian models of object perception. *Current Opinion in Neurobiology*, *13*(2), 150–158.

Khatib, O. (1987). A unified approach to motion and force control of robot manipulators: The operational space formulation. *IEEE Transactions on Robotics and Automation*, *3*(1), 43–53.

Kimura, T., Haggard, P. & Gomi, H. (2006). Transcranial magnetic stimulation over sensorimotor cortex disrupts anticipatory reflex gain modulation for skilled action. *Journal of Neuroscience*, *26*(36), 9272–9281.

Kirsch, R. F., Boskov, D. & Rymer, W. Z. (1994). Muscle stiffness during transient and continuous movements of cat muscle: Perturbation characteristics and physiological relevance. *IEEE Transactions on Bio-Medical Engineering*, *41*(8), 758–770.

Klassen, J., Tong, C. & Flanagan, J. R. (2005). Learning and recall of incremental kinematic and dynamic sensorimotor transformations. *Experimental Brain Research*, *164*(2), 250–259.

Kleim, J. A. & Jones, T. A. (2008). Principles of experience-dependent plasticity: Implication for re-habilitation after brain damage. *Journal of Speech, Language, and Hearing Research*, *51*(1), S225–239.

Kluzik, J., Diedrichsen, J., Shadmehr, R. & Bastian, A. J. (2008). Reach adaptation: What determines whether we learn an internal model of the tool or adapt the model of our arm? *Journal of Neurophysiology, 100*(3), 1455–1464.

Knill, D. C., Bondada, A. & Chhabra, M. (2011). Flexible, task-dependent use of sensory feedback to control hand movements. *Journal of Neuroscience, 31*(4), 1219–1237.

Knill, D. C. & Saunders, J. A. J. (2003). Do humans optimally integrate stereo and texture information for judgments of surface slant? *Vision Research, 43*(24), 2539–2558.

Kodl, J., Ganesh, G., & Burdet, E. (2011). CNS stochastically selects motor plan from extrinsic and intrinsic constraints. *PLoS ONE, 6*(9), e24229. doi:10.1371/journal.pone.0024229.

Kojima, Y., Iwamoto, Y. & Yoshida, K. (2004). Memory of learning facilitates saccadic adaptation in the monkey. *Journal of Neuroscience, 24*(34), 7531–7539.

Körding, K. P., Beierholm, U., Ma, W. J., Quartz, S., Tenenbaum, J. B. & Shams, L. (2007). Causal inference in multisensory perception. *PloS One, 2*(9), e943. doi:10.1371/journal.pone.0000943

Körding, K. P., Ku, S.-P. & Wolpert, D. M. (2004). Bayesian integration in force estimation. *Journal of Neurophysiology, 92*(5), 3161–3165.

Körding, K. P. & Wolpert, D. M. (2004). Bayesian integration in sensorimotor learning. *Nature, 427*(6971), 244–247.

Körding, K. P. & Wolpert, D. M. (2006). Bayesian decision theory in sensorimotor control. *Trends in Cognitive Sciences, 10*(7), 319–326.

Krakauer, J. W., Ghilardi, M. F. & Ghez, C. (1999). Independent learning of internal models for kinematic and dynamic control of reaching. *Nature Neuroscience, 2*(11), 1026–1031.

Krutky, M. A., Ravichandran, V. J., Trumbower, R. D. & Perreault, E. J. (2010). Interactions between limb and environmental mechanics influence stretch reflex sensitivity in the human arm. *Journal of Neurophysiology, 103*(1), 429–440.

Kurata, K. (2007). Laterality of movement-related activity reflects transformation of coordinates in ventral premotor cortex and primary motor cortex of monkeys. *Journal of Neurophysiology, 98*(4), 2008–2021.

Kurata, K. & Hoffman, D. S. (1994). Differential effects of muscimol microinjection into dorsal and ventral aspects of the premotor cortex of monkeys. *Journal of Neurophysiology, 71*(3), 1151–1164.

Kurtzer, I. L., Pruszynski, J. A. & Scott, S. H. (2008). Long-latency reflexes of the human arm reflect an internal model of limb dynamics. *Current Biology, 18*(6), 449–453.

Kurtzer, I., Pruszynski, J. A. & Scott, S. H. (2009). Long-latency responses during reaching account for the mechanical interaction between the shoulder and elbow joints. *Journal of Neurophysiology, 102*(5), 3004–3015.

Kutch, J. J. & Valero-Cuevas, F. J. (2012). Challenges and new approaches to proving the existence of muscle synergies of neural origin. *PLoS Computational Biology, 8*(5), e1002434. doi:10.1371/journal.pcbi.1002434.

Kwakkel, G., Kollen, B. J. & Krebs, H. I. (2008). Effects of robot-assisted therapy on upper limb recovery after stroke: A systematic review. *Neurorehabilitation and Neural Repair, 22*(2), 111–121.

Lackner, J. R. & DiZio, P. (1994). Rapid adaptation to coriolis-force perturbations of arm trajectory. *Journal of Neurophysiology, 72*(1), 299–313.

Lacquaniti, F., Borghese, N. A. & Carrozzo, M. (1991). Transient reversal of the stretch reflex in human arm muscles. *Journal of Neurophysiology, 66*(3), 939–954.

Lacquaniti, F., Carrozzo, M. & Borghese, N. A. (1993). Time-varying mechanical behavior of multijointed arm in man. *Journal of Neurophysiology, 69*(5), 1443–1464.

Lacquaniti, F. & Maioli, C. (1987). Anticipatory and reflex coactivation of antagonist muscles in catching. *Brain Research, 406*(1–2), 373–378.

Lacquaniti, F. & Maioli, C. (1989a). The role of preparation in tuning anticipatory and reflex responses during catching. *Journal of Neuroscience, 9*(1), 134–148.

Lacquaniti, F. & Maioli, C. (1989b). Adaptation to suppression of visual information during catching. *Journal of Neuroscience, 9*(1), 149–159.

Lacquaniti, F. & Soechting, J. F. (1986a). EMG responses to load perturbations of the upper limb: Effect of dynamic coupling between shoulder and elbow motion. *Experimental Brain Research*, *61*(3), 482–496.

Lacquaniti, F. & Soechting, J. F. (1986b). Responses of mono- and bi-articular muscles to load perturbations of the human arm. *Experimental Brain Research*, *65*(1), 135–144.

Lam, P., Hebert, D., Boger, J., Lacheray, H., Gardner, D., Apkarian, J., et al. (2008). A haptic-robotic platform for upper-limb reaching stroke therapy: Preliminary design and evaluation results. *Journal of Neuroengineering and Rehabilitation*, *5*(15). doi:10.1186/1743-0003-5-15.

Lambercy, O., Dovat, L., Yun, H., Wee, S. K., Kuah, C., Chua, K., Gassert, R., Milner, T. E., Teo, C. L. & Burdet, E. (2011). Robot-assisted rehabilitation of grasp and pronation/supination. *Journal of Neuroengineering and Rehabilitation*, *8*(63), doi:10.1186/1743-0003-8-63.

Lametti, D. R., Houle, G. & Ostry, D. J. (2007). Control of movement variability and the regulation of limb impedance. *Journal of Neurophysiology*, *98*(6), 3516–3524.

Lametti, D. R. & Ostry, D. J. (2010). Postural constraints on movement variability. *Journal of Neurophysiology*, *104*(2), 1061–1067.

Larsson, L., Grimby, G. & Karlsson, J. (1979). Muscle strength and speed of movement in relation to age and muscle morphology. *Journal of Applied Physiology*, *46*(3), 451–456.

Latash, M. L., Scholz, J. P. & Schöner, G. (2007). Toward a new theory of motor synergies. *Motor Control*, *11*(3), 276–308.

Lee, J.-Y. & Schweighofer, N. (2009). Dual adaptation supports a parallel architecture of motor memory. *Journal of Neuroscience*, *29*(33), 10396–10404.

Lee, R. G. & Tatton, W. G. (1982). Long latency reflexes to imposed displacements of the human wrist: Dependence on duration of movement. *Experimental Brain Research*, *45*(1–2), 207–216.

Leger, A. B. & Milner, T. E. (2000). The effect of eccentric exercise on intrinsic and reflex stiffness in the human hand. *Clinical Biomechanics*, *15*(8), 574–582.

Li, C. S., Padoa-Schioppa, C. & Bizzi, E. (2001). Neuronal correlates of motor performance and motor learning in the primary motor cortex of monkeys adapting to an external force field. *Neuron*, *30*(2), 593–607.

Lieber, R. L. & Ward, S. R. (2011). Skeletal muscle design to meet functional demands. *Philosophical Transactions of the Royal Society of London. Series B, Biological Sciences*, *366*(1570), 1466–1476.

Lindle, R. S., Metter, E. J., Lynch, N. A., Fleg, J. L., Fozard, J. L., Tobin, J., et al. (1997). Age and gender comparisons of muscle strength in 654 women and men aged 20–93 yr. *Journal of Applied Physiology*, *83*(5), 1581–1587.

Liu, D. & Todorov, E. (2007). Evidence for the flexible sensorimotor strategies predicted by optimal feedback control. *Journal of Neuroscience*, *27*(35), 9354–9368.

Maganaris, C. N. & Paul, J. P. (2002). Tensile properties of the in vivo human gastrocnemius tendon. *Journal of Biomechanics*, *35*(12), 1639–1646.

Majdandzic, J., Bekkering, H., van Schie, H. T. & Toni, I. (2009). Movement-specific repetition suppression in ventral and dorsal premotor cortex during action observation. *Cerebral Cortex*, *19*(11), 2736–2745.

Malamud, J. G., Godt, R. G. & Nichols, T. R. (1996). Relationship between short-range stiffness and yielding in type-identified, chemically skinned muscle fibers from the cat triceps surae muscles. *Journal of Neurophysiology*, *76*(4), 2280–2289.

Marr, D. (1969). A theory of cerebellar cortex. *Journal of Physiology*, *202*(2), 437–470.

Matthews, P. B. C. (1991). The human stretch reflex and the motor cortex. *Trends in Neurosciences*, *14*(3), 87–91.

McGurk, H., & MacDonald, J. (1976). Hearing lips and seeing voices. *Nature*, *264*(5588), 746–748.

Mehta, B., & Schaal, S. (2002). Forward models in visuomotor control. *Journal of Neurophysiology*, *88*(2), 942–953.

Melendez-Calderon, A. (2011). Investigating sensory-motor interactions to shape rehabilitation. Ph.D. Thesis, Imperial College of Science, Technology and Medicine.

Melendez-Calderon, A., Masia, L., Gassert, R., Sandini, G., & Burdet, E. (2011). Force field adaptation can be learned using vision in the absence of proprioceptive error. *IEEE Transactions on Neural Systems and Rehabilitation Engineering, 19*(3), 298–306.

Menzel, P. & d'Alusio, F. (2000). *Robo sapiens: Evolution of a new species.* Cambridge, MA: MIT Press.

Miall, R. C., Christensen, L. O. D., Cain, O. & Stanley, J. (2007). Disruption of state estimation in the human lateral cerebellum. *PLoS Biology, 5*(11), e316. doi:10.1371/journal.pbio.0050316.

Miall, R. C. & Reckess, G. Z. (2002). The cerebellum and the timing of coordinated eye and hand tracking. *Brain and Cognition, 48*(1), 212–226.

Miall, R. C., Weir, D. J. & Stein, J. F. (1993). Intermittency in human manual tracking tasks. *Journal of Motor Behavior, 25*(1), 53–63.

Milner, T. E. (1986a). Judgment and control of velocity in rapid voluntary movements. *Experimental Brain Research, 62*(1), 99–110.

Milner, T. E. (1986b). Controlling velocity in rapid movements. *Journal of Motor Behavior, 18*(2), 147–161.

Milner, T. E. (1992). A model for the generation of movements requiring endpoint precision. *Neuroscience, 49*(2), 487–496.

Milner, T. E. (1993). Dependence of elbow viscoelastic behavior on speed and loading in voluntary movements. *Experimental Brain Research, 93*(1), 177–180.

Milner, T. E. (2002). Contribution of geometry and joint stiffness to mechanical stability of the human arm. *Experimental Brain Research, 143*(4), 515–519.

Milner, T. E. (2004). Accuracy of internal dynamics models in limb movements depends on stability. *Experimental Brain Research, 159*(2) 172–184.

Milner, T. E. (2009). Impedance control. In M. D. Binder, N. Hirokawa, & U. Windhorst (Eds.), *Encyclopedia of neuroscience* (pp. 1929–1934). Berlin: Springer.

Milner, T. E. & Cloutier, C. (1993). Compensation for mechanically unstable loading in voluntary wrist movement. *Experimental Brain Research, 94*(3), 522–532.

Milner, T. E., Cloutier, C., Leger, A. B. & Franklin, D. W. (1995). Inability to maximally activate muscles during co-contraction and the effect on joint stiffness. *Experimental Brain Research, 107*(2), 293–305.

Milner, T. E., Dugas, C., Picard, N. & Smith, A. M. (1991). Cutaneous afferent activity in the median nerve during grasping in the primate. *Brain Research, 548*(1–2), 228–241.

Milner, T. E. & Franklin, D. W. (2005). Impedance control and internal model use during the initial stage of adaptation to novel dynamics in humans. *Journal of Physiology, 567*(Pt 2), 651–664.

Milner, T. E. & Ijaz, M. M. (1990). The effect of accuracy constraints on three-dimensional movement kinematics. *Neuroscience, 35*(2), 365–374.

Milner, T. E., Lai, E. J. & Hodgson, A. J. (2009). Modulation of arm stiffness in relation to instability at the beginning or the end of goal-directed movements. *Motor Control, 13*(4), 454–470.

Miyazaki, M., Nozaki, D. & Nakajima, Y. (2005). Testing Bayesian models of human coincidence timing. *Journal of Neurophysiology, 94*(1), 395–399.

Moberg, E. (1983). The role of cutaneous afferents in position sense, kinesthesia, and motor function of the hand. *Brain, 106*(Pt. 1), 1–19.

Morasso, P. (2011). "Brute force" vs. "gentle taps" in the control of unstable loads. *Journal of Physiology, 589*(Pt. 3), 459–460.

Mussa-Ivaldi, F. A. (1988). Do neurons in the motor cortex encode movement direction? An alternative hypothesis. *Neuroscience Letters, 91*(1), 106–111.

Mussa-Ivaldi, F. A. & Giszter, S. F. (1992). Vector field approximation: A computational paradigm for motor control and learning. *Biological Cybernetics, 67*(6), 491–500.

Mussa-Ivaldi, F. A. & Hogan, N. (1991). Integrable solutions of kinematic redundancy via impedance control. *International Journal of Robotics Research, 10*(5), 481–491.

Mussa-Ivaldi, F. A., Hogan, N. & Bizzi, E. (1985). Neural, mechanical, and geometric factors subserving arm posture in humans. *Journal of Neuroscience, 5*(10), 2732–2743.

Nachev, P., Kennard, C. & Husain, M. (2008). Functional role of the supplementary and pre-supplementary motor areas. *Nature Reviews. Neuroscience, 9*(11), 856–869.

Nader, K. & Hardt, O. (2009). A single standard for memory: The case for reconsolidation. *Nature Reviews. Neuroscience, 10*(3), 224–234.

Nanayakkara, T. & Shadmehr, R. (2003). Saccade adaptation in response to altered arm dynamics. *Journal of Neurophysiology, 90*(6), 4016–4021.

Nichols, T. R. & Houk, J. C. (1976). Improvement in linearity and regulation of stiffness that results from actions of stretch reflex. *Journal of Neurophysiology, 39*(1), 119–142.

Norris, A. H., Shock, N. W. & Wagman, I. H. (1953). Age changes in the maximum conduction velocity of motor fibers of human ulnar nerves. *Journal of Applied Physiology, 5*(10), 589–593.

Nozaki, D., Kurtzer, I. & Scott, S. H. (2006). Limited transfer of learning between unimanual and bimanual skills within the same limb. *Nature Neuroscience, 9*(11), 1364–1366.

Nyitrai, M. & Geeves, M. A. (2004). Adenosine diphosphate and strain sensitivity in myosin motors. *Philosophical Transactions of the Royal Society of London. Series B, Biological Sciences, 359*(1452), 1867–1877.

Øksendal, B. (1998). *Stochastic differential equations*. Berlin: Springer Verlag.

Orban de Xivry, J.-J., Criscimagna-Hemminger, S. E. & Shadmehr, R. (2011). Contributions of the motor cortex to adaptive control of reaching depend on perturbation schedule. *Cerebral Cortex, 21*(7), 1475–1484.

Osu, R., Burdet, E., Franklin, D. W., Milner, T. E. & Kawato, M. (2003). Different mechanisms involved in adaptation to stable and unstable dynamics. *Journal of Neurophysiology, 90*(5), 3255–3269.

Osu, R., Franklin, D. W., Kato, H., Gomi, H., Domen, K., Yoshioka, T., et al. (2002). Short- and long-term changes in joint co-contraction associated with motor learning as revealed from surface EMG. *Journal of Neurophysiology, 88*(2), 991–1004.

Osu, R., Kamimura, N., Iwasaki, H., Nakano, E., Harris, C. M., Wada, Y., et al. (2004). Optimal impedance control for task achievement in the presence of signal-dependent noise. *Journal of Neurophysiology, 92*(2), 1199–1215.

O'Sullivan, I., Burdet, E. and Diedrichsen, J. (2009). Dissociating variability and effort as determinants of coordination. *Plos Computational Biology, 5*(4): e1000345.

Overduin, S. A., Richardson, A. G., Lane, C. E., Bizzi, E. & Press, D. Z. (2006). Intermittent practice facilitates stable motor memories. *Journal of Neuroscience, 26*(46), 11888–11892.

Palazzolo, J. J., Ferraro, M., Krebs, H. I., Lynch, D., Volpe, B. T. & Hogan, N. (2007). Stochastic estimation of arm mechanical impedance during robotic stroke rehabilitation. *IEEE Transactions on Neural Systems and Rehabilitation Engineering, 15*(1), 94–103.

Papaxanthis, C., Pozzo, T. & Schieppati, M. (2003). Trajectories of arm pointing movements on the sagittal plane vary with both direction and speed. *Experimental Brain Research, 148*(4), 498–503.

Partridge, L. D. (1965). Modifications of neural output signals by muscles: A frequency response study. *Journal of Applied Physiology, 20*, 150–156.

Patton, J. L. & Huang, F. C. (2012). Error augmentation and the role of sensory feedback. In V. Dietz, T. Nef, & W. Z. Rymer (Eds.), *Neurorehabilitation technology* (pp. 73–86). London: Springer.

Patton, J. L., Kovic, M. & Mussa-Ivaldi, F. A. (2006). Custom-designed haptic training for restoring reaching ability to individuals with poststroke hemiparesis. *Journal of Rehabilitation Research and Development, 43*(5), 643–656.

Paulignan, Y., Jeannerod, M., MacKenzie, C. & Marteniuk, R. (1991). Selective perturbation of visual input during prehension movements. *Experimental Brain Research, 87*(2), 407–420.

Perreault, E. J., Chen, K., Trumbower, R. D. & Lewis, G. (2008). Interactions with compliant loads alter stretch reflex gains but not intermuscular coordination. *Journal of Neurophysiology, 99*(5), 2101–2113.

Perreault, E. J., Kirsch, R. F. & Crago, P. E. (2002). Voluntary control of static endpoint stiffness during force regulation tasks. *Journal of Neurophysiology, 87*(6), 2808–2816.

Perreault, E. J., Kirsch, R. F. & Crago, P. E. (2004). Multijoint dynamics and postural stability of the human arm. *Experimental Brain Research, 157*(4), 507–517.

Pesaran, B., Nelson, M. J. & Andersen, R. A. (2006). Dorsal premotor neurons encode the relative position of the hand, eye, and goal during reach planning. *Neuron*, *51*(1), 125–134.

Prablanc, C. & Martin, O. (1992). Automatic control during hand reaching at undetected two-dimensional target displacements. *Journal of Neurophysiology*, *67*(2), 455–469.

Prablanc, C., Pélisson, D. & Goodale, M. A. (1986). Visual control of reaching movements without vision of the limb. *Experimental Brain Research*, *62*(2), 293–302.

Prange, G. B., Jannink, M. J., Groothuis-Oudshoorn, C. G. M., Hermens, H. J. & Ijzerman, M. J. (2006). Systematic review of the effect of robot-aided therapy on recovery of the hemiparetic arm after stroke. *Journal of Rehabilitation Research and Development*, *43*(2), 171–184.

Prochazka, A. (2011). Proprioceptive feedback and movement regulation. *Comprehensive Physiology*, 89–127. doi:10.1002/cphy.cp120103.

Prochazka, A., Gillard, D. & Bennett, D. J. (1997). Positive force feedback control of muscles. *Journal of Neurophysiology*, *77*(6), 3226–3236.

Proske, U. & Gregory, J. E. (2002). Signalling properties of muscle spindles and tendon organs. *Advances in Experimental Medicine and Biology*, *508*, 5–12.

Pruszynski, J. A., Kurtzer, I., Lillicrap, T. P. & Scott, S. H. (2009). Temporal evolution of "automatic gain-scaling." *Journal of Neurophysiology*, *102*(2), 992–1003.

Pruszynski, J. A., Kurtzer, I., Nashed, J. Y., Omrani, M., Brouwer, B. & Scott, S. H. (2011). Primary motor cortex underlies multi-joint integration for fast feedback control. *Nature*, *478*(7369), 387–390.

Pruszynski, J. A. & Scott, S. H. (2012). Optimal feedback control and the long-latency stretch response. *Experimental Brain Research*, *218*(3), 341–359.

Purves, D., Augustine, G. J., Fitzpatrick, D., Hall, W. C., LaMantia, A.-S. & White, L. E. (2012). *Neuroscience*. Sunderland, MA: Sinauer Associates.

Rack, P. M. H. (2011). Limitations of somatosensory feedback in control of posture and movement. *Comprehensive Physiology*, 229–256. doi:10.1002/cphy.cp120207.

Rancourt, D. & Hogan, N. (2001). Stability in force-production tasks. *Journal of Motor Behavior*, *33*(2), 193–204.

Rebsamen, B., Guan, C., Zhang, H., Wang, C., Teo, C. L., Ang, M., et al. (2010). A brain controlled wheelchair to navigate in familiar environments. *IEEE Transactions on Neural Systems and Rehabilitation Engineering*, *18*(6), 590–598.

Reinkensmeyer, D. J., Schmit, B. D. & Rymer, W. Z. (1999). Assessment of active and passive restraint during guided reaching after chronic brain injury. *Annals of Biomedical Engineering*, *27*(6), 805–814.

Richardson, A. G., Overduin, S. A., Valero-Cabré, A., Padoa-Schioppa, C., Pascual-Leone, A., Bizzi, E. & Press, D. Z. (2006). Disruption of primary motor cortex before learning impairs memory of movement dynamics. *Journal of Neuroscience*, *26*(48), 12466–12470.

Rigoux, L. & Guigon, E. (2012). A model of reward- and effort-based optimal decision making and motor control. *PLoS Computational Biology*, *8*(10), e1002716. doi:10.1371/journal.pcbi.1002716.

Riley, K. F., Hobson, M. P. & Bence, S. J. (2006). *Mathematical methods for physics and engineering*. Cambridge: Cambridge University Press.

Rizolatti, G. & Craighero, L. (2004). The mirror-neuron system. *Annual Review of Neuroscience*, *27*, 169–192.

Roy, J. E. & Cullen, K. E. (2001). Selective processing of vestibular reafference during self-generated head motion. *Journal of Neuroscience*, *21*(6), 2131–2142.

Roy, J. E. & Cullen, K. E. (2004). Dissociating self-generated from passively applied head motion: Neural mechanisms in the vestibular nuclei. *Journal of Neuroscience*, *24*(9), 2102–2111.

Saijo, N., Murakami, I., Nishida, S. & Gomi, H. (2005). Large-field visual motion directly induces an involuntary rapid manual following response. *Journal of Neuroscience*, *25*(20), 4941–4951.

Sanner, R. M. & Kosha, M. (1999). A mathematical model of the adaptive control of human arm motions. *Biological Cybernetics*, *80*(5), 369–382.

Sanner, R. & Slotine, J. J.-E. (1992). Gaussian networks for direct adaptive control. *IEEE Transactions on Neural Networks*, *3*(6), 837–863.

Sarlegna, F., Blouin, J., Bresciani, J.-P., Bourdin, C., Vercher, J.-L. & Gauthier, G. M. (2003). Target and hand position information in the online control of goal-directed arm movements. *Experimental Brain Research*, *151*(4), 524–535.

Saunders, J. A. & Knill, D. C. (2003). Humans use continuous visual feedback from the hand to control fast reaching movements. *Experimental Brain Research*, *152*(3), 341–352.

Saunders, J. A. & Knill, D. C. (2004). Visual feedback control of hand movements. *Journal of Neuroscience*, *24*(13), 3223–3234.

Scangos, K. W. & Stuphorn, V. (2010). Medial frontal cortex motivates but does not control movement initiation in the countermanding task. *Journal of Neuroscience*, *30*(5), 1968–1982.

Scheidt, R. A., Dingwell, J. B. & Mussa-Ivaldi, F. A. (2001). Learning to move amid uncertainty. *Journal of Neurophysiology*, *86*(2), 971–985.

Scheidt, R. A., Conditt, M. A., Secco, E. L. & Mussa-Ivaldi, F. A. (2005). Interaction of visual and proprioceptive feedback during adaptation of human reaching movements. *Journal of Neurophysiology*, *93*(6), 3200–3213.

Scheidt, R. A., Reinkensmeyer, D. J., Conditt, M. A., Rymer, W. Z. & Mussa-Ivaldi, F. A. (2000). Persistence of motor adaptation during constrained, multi-joint, arm movements. *Journal of Neurophysiology*, *84*(2), 853–862.

Scholz, J. P. & Schöner, G. (1999). The uncontrolled manifold concept: Identifying control variables for a functional task. *Experimental Brain Research*, *126*(3), 289–306.

Scott, S. H. (2004). Optimal feedback control and the neural basis of volitional motor control. *Nature Reviews Neuroscience*, *5*(7), 532–546.

Selen, L. P. J., Beek, P. J. & van Dieën, J. H. (2005). Can co-activation reduce kinematic variability? A simulation study. *Biological Cybernetics*, *93*(5), 373–381.

Selen, L. P. J., Beek, P. J. & van Dieën, J. H. (2006). Impedance is modulated to meet accuracy demands during goal-directed arm movements. *Experimental Brain Research*, *172*(1), 129–138.

Selen, L. P. J., Franklin, D. W. & Wolpert, D. M. (2009). Impedance control reduces instability that arises from motor noise. *Journal of Neuroscience*, *29*(40), 12606–12616.

Selen, L. P. J., van Dieën, J. H. & Beek, P. J. (2006). Impedance modulation and feedback corrections in tracking targets of variable size and frequency. *Journal of Neurophysiology*, *96*(5), 2750–2759.

Sentis, L. & Khatib, O. (2005). Synthesis of whole-body behaviors through hierarchical control of behavioral primitives. *International Journal of Humanoid Robotics*, *2*(4), 505–518.

Shadmehr, R. & Mussa-Ivaldi, F. A. (1994). Adaptive representation of dynamics during learning of a motor task. *Journal of Neuroscience*, *74*(5), 3208–3224.

Shadmehr, R., Mussa-Ivaldi, F. A. & Bizzi, E. (1993). Postural force fields of the human arm and their role in generating multi-joint movements. *Journal of Neuroscience*, *13*(1), 45–62.

Shadmehr, R. & Brashers-Krug, T. (1997). Functional stages in the formation of human long-term motor memory. *Journal of Neuroscience*, *17*(1), 409–419.

Shadmehr, R. & Wise, S. P. (2005). *The computational neurobiology of reaching and pointing: A foundation for motor learning*. Cambridge, MA: MIT Press.

Shemmell, J., An, J. H. & Perreault, E. J. (2009). The differential role of motor cortex in stretch reflex modulation induced by changes in environmental mechanics and verbal instruction. *Journal of Neuroscience*, *29*(42), 13255–13263.

Shergill, S. S., Bays, P. M., Frith, C. D. & Wolpert, D. M. (2003). Two eyes for an eye: The neuroscience of force escalation. *Science*, *301*(5630), 187.

Shimansky, Y. P., Kang, T. & He, J. (2004). A novel model of motor learning capable of developing an optimal movement control law online from scratch. *Biological Cybernetics*, *90*(2), 133–145.

Siciliano, B., Sciavicco, L., Villani, L. & Oriolo, G. (2009). *Robotics: Modelling, planning and control*. London: Springer.

Simon, D. (2006). *Optimal state estimation: Kalman, H_∞, and nonlinear approaches*. Hoboken, NJ: Wiley.

Sinkjaer, T., Toft, E., Andreassen, S. & Hornemann, B. C. (1988). Muscle stiffness in human ankle dorsiflexors: Intrinsic and reflex components. *Journal of Neurophysiology*, *60*(3), 1110–1121.

Slifkin, A. B. & Newell, K. M. (1999). Noise, information transmission, and force variability. *Journal of Experimental Psychology*, *25*(3), 837–851.

Slotine, J.-J. E. & Li, W. (1991). *Applied nonlinear control*. Englewood Cliffs, NJ: Prentice Hall.

Smith, M. A., Ghazizadeh, A. & Shadmehr, R. (2006). Interacting adaptive processes with different timescales underlie short-term motor learning. *PLoS Biology*, *4*(6), e179. doi:10.1371/journal.pbio.0040179.sd001.

Sober, S. J. & Sabes, P. N. (2005). Flexible strategies for sensory integration during motor planning. *Nature Neuroscience*, *8*(4), 490–497.

Soechting, J. F. & Lacquaniti, F. (1983). Modification of trajectory of a pointing movement in response to a change in target location. *Journal of Neurophysiology*, *49*(2), 548–564.

Soechting, J. F. & Lacquaniti, F. (1988). Quantitative evaluation of the electromyographic responses to multidirectional load perturbations of the human arm. *Journal of Neurophysiology*, *59*(4), 1296–1313.

Spong, M. W., Hutchinson, S. & Vidyasagar, M. (2006). *Robot modeling and control*. Hoboken, NJ: Wiley.

Stengel, R. F. (1994). *Optimal control and estimation*. New York: Dover Publications.

Stevens, J. R., Rosati, A. G., Ross, K. R. & Hauser, M. D. (2005). Will travel for food: Spatial discounting in two new world monkeys. *Current Biology*, *15*(20), 1855–1860.

Sugisaki, N., Kawakami, Y., Kanehisa, H. & Fukunaga, T. (2011). Effect of muscle contraction levels on the force-length relationship of the human Achilles tendon during lengthening of the triceps surae muscle-tendon unit. *Journal of Biomechanics*, *44*(11), 2168–2171.

Takahashi, C. D., Der-Yeghiaian, L. L. V., Motiwala, R. R. & Cramer, S. C. (2008). Robot-based hand motor therapy after stroke. *Brain*, *131*(2), 425–437.

Tansey, K. E. & Botterman, B. R. (1996). Activation of type-identified motor units during centrally evoked contractions in the cat medial gastrocnemius muscle. I. Motor-unit recruitment. *Journal of Neurophysiology*, *75*(1), 26–37.

Tee, K. P., Burdet, E., Chew, C. M. & Milner, T. E. (2004). A model of force and impedance in human arm movements. *Biological Cybernetics*, *90*(5), 368–375.

Tee, K. P., Franklin, D. W., Kawato, M., Milner, T. E. & Burdet, E. (2010). Concurrent adaptation of force and impedance in the redundant muscle system. *Biological Cybernetics*, *102*(1), 31–44.

Theodorou, E., Buchli, J. & Schaal, S. (2010). A generalized path integral control approach to reinforcement learning. *Journal of Machine Learning Research*, 11, 3137–3181.

Thoroughman, K. A. & Shadmehr, R. (1999). Electromyographic correlates of learning an internal model of reaching movements. *Journal of Neuroscience*, *19*(19), 8573–8588.

Thoroughman, K. A. & Shadmehr, R. (2000). Learning of action through adaptive combination of motor primitives. *Nature*, *407*(6805), 742–747.

Thoroughman, K. A., & Taylor, J. A. (2005). Rapid reshaping of human motor generalization. *Journal of Neuroscience*, *25*(39), 8948–8953.

Todorov, E. (2004). Optimality principles in sensorimotor control. *Nature Neuroscience*, *7*(9), 907–915.

Todorov, E. (2007). Optimal control theory. In K. Doya, S. Ishi, A. Pouget, & R. P. N. Rao (Eds.), *Bayesian brain: Probabilistic approaches to neural coding* (pp. 269–298). Cambridge, MA: MIT Press.

Todorov, E. & Jordan, M. I. (2002). Optimal feedback control as a theory of motor coordination. *Nature Neuroscience*, *5*(11), 1226–1235.

Tong, C., Wolpert, D. M. & Flanagan, J. R. (2002). Kinematics and dynamics are not represented independently in motor working memory: Evidence from an interference study. *Journal of Neuroscience*, *22*(3), 1108–1113.

Trumbower, R. D., Krutky, M. A., Yang, B.-S. & Perreault, E. J. (2009). Use of self-selected postures to regulate multi-joint stiffness during unconstrained tasks. *PLoS ONE*, *4*(5), e5411. doi:10.1371/journal.pone.0005411.

Tseng, Y. W., Diedrichsen, J., Krakauer, J. W., Shadmehr, R. & Bastian, A. J. (2007). Sensory prediction errors drive cerebellum-dependent adaptation of reaching. *Journal of Neurophysiology*, *98*(1), 54–62.

Turner, R. S. & Desmurget, M. (2010). Basal ganglia contributions to motor control: A vigorous tutor. *Current Opinion in Neurobiology*, *20*(6), 704–716.

Tweed, D. & Vilis, T. (1990). Geometric relations of eye position and velocity vectors during saccades. *Vision Research*, *30*(1), 111–127.

Uno, Y., Kawato, M. & Suzuki, R. (1989). Formation and control of optimal trajectory in human multijoint arm movement. *Biological Cybernetics*, *61*(2), 89–101.

Vahdat, S., Darainy, M., Milner, T. E. & Ostry, D. J. (2011). Functionally specific changes in resting-state sensorimotor networks after motor learning. *Journal of Neuroscience*, *31*(47), 16907–16915.

van Beers, R. J., Haggard, P. & Wolpert, D. M. (2004). The role of execution noise in movement variability. *Journal of Neurophysiology*, *91*(2), 1050–1063.

van Beers, R. J., Sittig, A. C. & Denier van der Gon, J. J. (1996). How humans combine simultaneous proprioceptive and visual position information. *Experimental Brain Research*, *111*(2), 253–261.

van Beers, R., Sittig, A. & Denier van der Gon, J. J. (1998). The precision of proprioceptive position sense. *Experimental Brain Research*, *122*(4), 367–377.

van Beers, R. J., Wolpert, D. M. & Haggard, P. (2002). When feeling is more important than seeing in sensorimotor adaptation. *Current Biology*, *12*(10), 834–837.

van Sonderen, J. F., Denier van der Gon, J. J. & Gielen, C. C. (1988). Conditions determining early modification of motor programmes in response to changes in target location. *Experimental Brain Research*, *71*(2), 320–328.

Vindras, P., Desmurget, M., Prablanc, C. & Viviani, P. (1998). Pointing errors reflect biases in the perception of the initial hand position. *Journal of Neurophysiology*, *79*(6), 3290–3294.

Vindras, P. & Viviani, P. (1998). Frames of reference and control parameters in visuomanual pointing. *Journal of Experimental Psychology*, *24*(2), 569–591.

von Hofsten, C. (1991). Structuring of early reaching movements: A longitudinal study. *Journal of Motor Behavior*, *23*(4), 280–292.

von Hofsten, C. & Rönnqvist, L. (1993). The structuring of neonatal arm movements. *Child Development*, *64*(4), 1046–1057.

Wade, D. T., Langton-Hewer, R., Wood, V. A., Skilbeck, C. E. & Ismail, H. M. (1983). The hemiplegic arm after stroke: Measurement and recovery. *Journal of Neurology, Neurosurgery, and Psychiatry*, *46*(6), 521–524.

Wadman, W. J., Denier van der Gon, J. J., Geuze, R. H. & Mol, C. R. (1979). Control of fast goal-directed arm movement. *Journal of Human Movement Studies*, *5*, 3–17.

Wang, F., Poston, T., Teo, C. L., Lim, K. L. & Burdet, E. (2004), Multisensory learning cues using analytical collision detection between a needle and a tube. *Proc Int Symp on Haptic Interfaces for Virtual Environment and Teleoperator Systems (HAPTICS)*, 339–346.

Wann, J. P. & Ibrahim, S. F. (1992). Does limb proprioception drift? *Experimental Brain Research*, *91*(1), 162–166.

Wei, Y., Bajaj, P., Scheidt, R. A. & Patton, J. (2005). Visual error augmentation for enhancing motor learning and rehabilitative relearning. *Proc IEEE International Conference on Rehabilitation Robotics (ICORR)*, 505–510.

Weiss, P. L., Hunter, I. W. & Kearney, R. E. (1988). Human ankle joint stiffness over the full range of muscle activation levels. *Journal of Biomechanics*, *21*(7), 539–544.

Winstein, C. J. & Wolf, S. L. (2008). Task-oriented training to promote upper extremity recovery. In J. Stein, R. L. Harvey, R. F. Macko, C. J. Winstein, & R. D. Zorowitz (Eds.), *Stroke recovery and rehabilitation* (pp. 267–290). New York: Demos Medical Publishing.

Winter, D. A. (2009). *Biomechanics and motor control of human movement*. Hoboken, NJ: Wiley.

Wolbrecht, E. T., Chan, V., Reinkensmeyer, D. J. & Bobrow, J. E. (2008). Optimizing compliant, model-based robotic assistance to promote neurorehabilitation. *IEEE Transactions on Neural Systems and Rehabilitation Engineering*, *16*(3), 286–297.

Wolpert, D. M. & Flanagan, J. R. (2001). Motor prediction. *Current Biology*, *11*(18), R729–R732.

Wolpert, D. M., Ghahramani, Z. & Jordan, M. I. (1995). An internal model for sensorimotor integration. *Science*, *269*(5232), 1880–1882.

Wolpert, D. M. & Kawato, M. (1998). Multiple paired forward and inverse models for motor control. *Neural Networks*, *11*(7–8), 1317–1329.

Wolpert, D. M. & Miall, R. C. (1996). Forward models for physiological motor control. *Neural Networks*, *9*(8), 1265–1279.

Wolpert, D. M., Miall, R. C. & Kawato, M. (1998). Internal models in the cerebellum. *Trends in Cognitive Sciences*, *2*(9), 338–347.

Won, J. & Hogan, N. (1995). Stability properties of human reaching movements. *Experimental Brain Research*, *107*(1), 125–136.

Xiao, J., Padoa-Schioppa, C. & Bizzi, E. (2006). Neuronal correlates of movement dynamics in the dorsal and ventral premotor area in the monkey. *Experimental Brain Research*, *168*(1–2), 106–119.

Yang, C., Ganesh, G., Haddadin, S., Parusel, S., Albu-Schaeffer, A. & Burdet, E. (2011). Human-like adaptation of force and impedance in stable and unstable interactions. *IEEE Transactions on Robotics*, *27*(5), 918–930.

Yeong, C. F., Melendez-Calderon, A., Gassert, R. & Burdet, E. (2009). ReachMAN: A personal robot to train reaching and manipulation, *Proc. IEEE/RSJ International Conference on Intelligent Robots and Systems (IROS)*, 4080–4085.

Zajac, F. E. (1993). Muscle coordination of movement: A perspective. *Journal of Biomechanics*, *26*(1), 109–124.

Zefran, M. & Kumar, V. (2002). A geometrical approach to the study of the Cartesian stiffness matrix. *Journal of Mechanical Design*, *124*(1), 30–38.

Index